Small Gas Engine Repair

fourth edition

Paul Dempsey

New York   Chicago   San Francisco   Athens   London
Madrid   Mexico City   Milan   New Delhi
Singapore   Sydney   Toronto

About the Author

Paul Dempsey is an experienced mechanic and the author of more than 30 technical books, including *How to Repair Briggs & Stratton Engines* and *Troubleshooting and Repairing Diesel Engines,* both published by McGraw-Hill.

Library of Congress Control Number: 2017937460

McGraw-Hill Education books are available at special quantity discounts to use as premiums and sales promotions or for use in corporate training programs. To contact a representative, please visit the Contact Us page at www.mhprofessional.com.

Small Gas Engine Repair, Fourth Edition

1 2 3 4 5 6 7 8 9   LCR   22 21 20 19 18 17

ISBN 978-1-259-86158-1
MHID 1-259-86158-9

This book is printed on acid-free paper.

Sponsoring Editor: Wendy Fuller
Editing Supervisor: Stephen M. Smith
Production Supervisor: Pamela A. Pelton
Acquisitions Coordinator: Lauren Rogers
Project Managers: Touseen Qadri and Poonam Bisht, MPS Limited
Copy Editor: Lucy Mullins
Proofreader: A. Nayyer Shamsi
Indexer: May Hasso
Art Director, Cover: Jeff Weeks
Composition: MPS Limited

Contents

Preface

This book describes how to repair lawnmowers that won't start and how to cope with balky weed eaters or home generators that shut down in the middle of a power outage. It also explores how small engines function. This background knowledge makes repairing easier and opens the mind to the beauty of things well constructed. The Honda Mini-4 is a monument to human ingenuity.

Repairing small engines is a lucrative business: Shops charge as much as $90 an hour for labor, which means that a tune-up on a riding mower—consisting of replacing spark plugs and filters and an oil change—costs nearly $200. Doing the work yourself, which is not difficult, involves an outlay of $50 for parts.

The initial chapter explains the basic concepts and establishes the vocabulary needed to buy parts and communicate with service technicians. Readers new to the technology will learn the difference between two- and four-stroke engines and hybrids that combine features of both. Exhaust emissions, ethanol fuels, and a dozen other topics are discussed, together with features to look for when shopping for a reliable, long-lasting engine.

The next chapter is an introduction to troubleshooting. Once the problem is identified, subsequent chapters describe how to make the repairs. Ignition, fuel, starting, and charging systems each have a chapter. The book concludes with detailed instructions on rebuilding two- and four-cycle engines—how much wear is tolerable and the ways to make a worn engine perform like new.

This new fourth edition of *Small Gas Engine Repair* has been updated to include information on the new generation of carburetors, electronic fuel injection, and stratified induction systems. The effort that went into writing the book will be worthwhile if readers find it helpful and informative.

Paul Dempsey

1

Basics

Four-cycle engines

Most small engines operate on the four-stroke cycle. As shown in Figure 1-1, this cycle consists of four events:

- Intake. As the piston falls on the downstroke, a mixture of fuel and air enters the cylinder past the open intake valve (A). The valve remains open until the piston nears its lower limit of movement, known as bottom dead center (bdc). The center is "dead" because the piston must come to a halt and reverse direction.
- Compression. The piston rounds bdc and begins to rise (B). Since both valves are closed, the cylinder is basically airtight. Cylinder pressure increases as the piston approaches top dead center (tdc). Up to a point determined by the onset of uncontrolled combustion, the more the air/fuel mixture is compressed prior to ignition, the greater the explosive force.
- Expansion or power. At about 20° of crankshaft rotation before top dead center (btdc), the spark plug fires. The electric arc ignites the air/fuel mixture and drives the piston down on the expansion stroke (C). This stroke is the whole purpose of the exercise.
- Exhaust. Near the end of the expansion stroke, just before the piston rounds bdc, the exhaust valve opens. Burnt gases blow down past the open valve. The piston then begins to climb on the exhaust stroke to force what gases remain in the cylinder out and into the atmosphere (D). The exhaust valve closes as the piston approaches tdc, and the intake valve opens in preparation for the advent of the intake stroke.

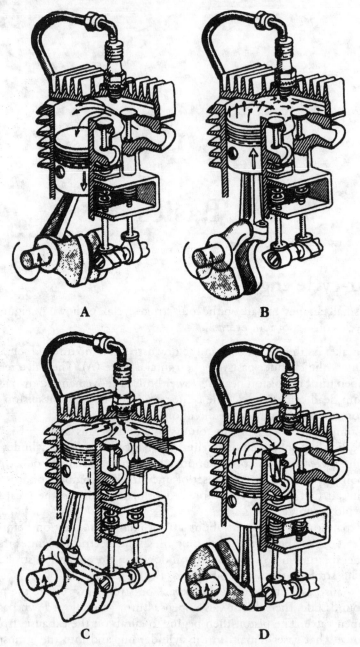

FIGURE 1-1. *The four-stroke cycle consists of intake (A), compression (B), spark and expansion (C), and exhaust (D).*

Four-cycle, or four-stroke (the terms are synonymous), engines require two full revolutions of the crankshaft to complete the operating cycle. The expansion stroke occurs during the second crankshaft revolution, which is a disadvantage. Ideally each downstroke of the piston should be a power stroke. On the other hand, a whole piston stroke is devoted to clearing the cylinder of exhaust gas residues. This, from the point of view of environmental protection, is an advantage, as we will see in the discussion of two-cycle engines that follows.

Valve configuration

Four-cycle engines can be recognized by the presence of intake and exhaust valves. Where these valves are located relative to the cylinder bore varies. Side-valve engines, also known as L-head or flathead engines, have their valves tucked alongside the cylinder (Fig. 1-2). The camshaft, driven at half crankshaft speed, bears directly on the underside of the valves. Flathead engines are anachronisms, dating back to the Wilson administration and earlier, but they have the virtue of simplicity. Briggs & Stratton turned them out by the millions.

Overhead valve (ohv), I-head, or valve-in-head engines are a much more modern concept (Fig. 1-3). In this configuration the valves mount facedown over the cylinder bore and operate through pushrods and pivoted rocker arms. The ohv configuration has several advantages, not the least of which

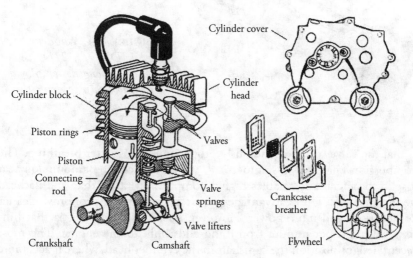

FIGURE 1-2. *Tecumseh side-valve, horizontal-shaft engine.*

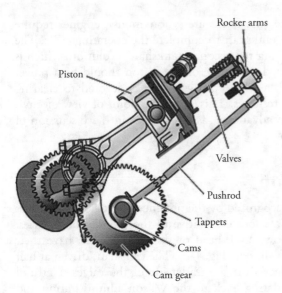

Rocker arms

Piston

Valves

Pushrod

Tappets

Cams

Cam gear

FIGURE 1-3. *Pushrods and rocker arms actuate the valves on ohv engines.* Subaru

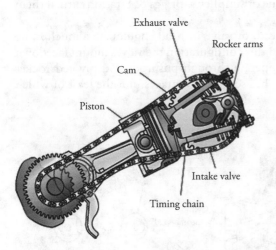

Exhaust valve

Rocker arms

Cam

Piston

Intake valve

Timing chain

FIGURE 1-4. *Subaru EX13D, 17D, 21D, and 27D ohc engines employ a timing chain to drive the camshaft.*

is that the valves can be generously proportioned for better breathing. The combustion chamber is symmetrical, which reduces the amount of heat lost to the metal and enables higher compression ratios than the flathead design permits. These advantages result in better fuel economy, cleaner combustion, and more power. However, pushrods create inertia that limits revolutions per minute (rpm) and adds bulk to the engine. The most recent designs position the camshaft in the cylinder head, where it acts more directly on the valves (Fig. 1-4). A toothed belt or a chain transfers motion to the cam from the crankshaft, and in several designs acts as a conveyor belt to carry oil to the valves. An overhead camshaft (ohc) gives precise control over valve motion, reduces rpm-limiting valve-gear inertia, and simplifies manufacturing. Suzuki says its ohc engines contain 30 percent fewer parts than comparable ohv engines.

Lubrication

With just a couple of exceptions, four-cycle engines lubricate from oil stored in the crankcase sump. How the oil is distributed over the moving parts varies. Side-valve engines with a horizontal crankshaft splash the oil about by means of an extension on the connecting rod; vertical-shaft engines use a camshaft-driven slinger for the same purpose (Fig. 1-5). Because side-valve camshafts live in the splash-drenched crankcase, they get plenty of lubrication. For ohv and ohc engines matters are complicated since some way must be found to move oil from the crankcase reservoir to the cylinder head and valve gear. Most of these engines incorporate an oil pump for topside delivery (Fig. 1-6), although as mentioned above, the camshaft drive can also be used for this purpose.

FIGURE 1-5. *Eaton-style eccentric-rotor pumps are a mark of quality.*

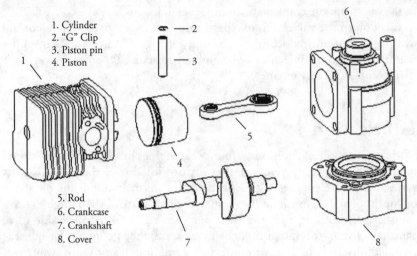

1. Cylinder
2. "G" Clip
3. Piston pin
4. Piston

5. Rod
6. Crankcase
7. Crankshaft
8. Cover

FIGURE 1-6. *A Tecumseh two-stroke engine, less carburetor, flywheel, ignition module, muffler, and shrouding. The one-piece cylinder barrel and cylinder head is typical of these engines.*

Two-cycle engines

Two-cycle, or two-stroke, engines share the same operational events— intake, compression, expansion, and exhaust—as four-cycle engines. The difference is that two-cycle engines telescope these events into two piston strokes, or one flywheel revolution. Every revolution of the crankshaft produces a power pulse, which means that these engines are theoretically twice as powerful as four strokes of the same displacement. The actual advantage ranges from 50 to 70 percent.

Two-cycle engines have two zones of compression—the cylinder bore and the crankcase—and three types of ports. The intake port delivers the fuel charge from the carburetor to the crankcase; the transfer port moves the fuel charge from the crankcase to the upper cylinder; and the exhaust port vents the cylinder to the atmosphere. The piston compresses air and fuel in the cylinder during the upstroke and in the crankcase on the downstroke. The piston also acts as a slide valve to open and close the exhaust and transfer ports and, on some engines, the intake port.

Piston-ported or third-port induction

As shown in Figure 1-7, engines built on the third-port pattern use the piston to control three ports. In drawing A, the piston has opened

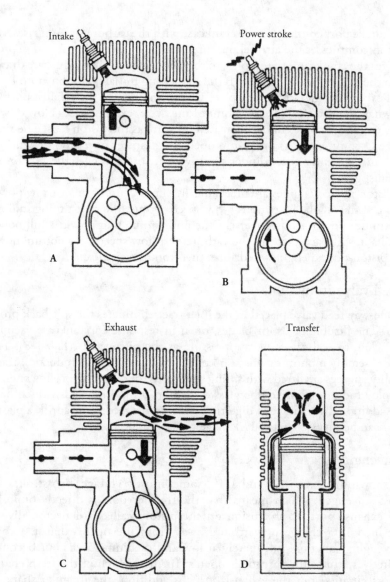

FIGURE 1-7. *Piston-ported two-strokes remain popular for light motorcycles and mopeds. Like other modern two-strokes, the engine is loop scavenged as shown in drawings C and D.* Used by permission of Walbro LLC

the intake port connecting the crankcase with the carburetor. As it rises, the piston compresses the air/fuel mixture in the cylinder above it and depressurizes the crankcase. In response to the low pressure, air and fuel are drawn into the crankcase. In drawing B the spark plug initiates combustion. Heat and expanding gases force the piston down to uncover the exhaust port C. In what is called the "release" point of the cycle, exhaust gases blow down. The crankcase comes under increased pressure as the piston falls. The final stage D commences as the piston opens the transfer ports and pressurized fuel and air enter the cylinder to drive out, or scavenge, the remaining exhaust gas.

Piston-ported engines are mechanically simple, but have a narrow power band. If the engine is set up for low-speed torque, the port timing inhibits maximum power. And conversely, the port timing that provides full power results in spit back through the carburetor at low speeds. It's not unknown for piston-ported engines to dribble fuel from the air cleaner.

Reed valve induction

A one-way reed valve placed in the inlet tract eliminates the spit-back problem. The flexible reeds, or blades, open in response to crankcase vacuum and close to seal off crankcase pressure as the piston falls (Fig. 1-8). The reed assembly mounts downstream of the carburetor in the induction tract. Utility engines are fitted with spring steel reeds for durability; racing engines employ more responsive fiberglass or carbon composite reeds, which have the additional advantage of digestibility. Should a reed come adrift, a plastic reed can be ingested without damage to the engine.

Scavenging

As shown in Figures 1-9 and 1-10, some, mostly vintage, two-strokes are cross-scavenged, which means that the transfer port is directly opposite the exhaust port. A protrusion on the piston, called a deflector, diverts the incoming charge upward and away from the open exhaust. Cross-scavenging functions adequately at low and medium speeds but becomes quite wasteful as the throttle is opened further. Much of the fresh charge "short circuits" out the exhaust port. In addition, the irregular shape of the piston creates hot spots that limit the allowable compression ratio and engine power.

Loop scavenging was the breakthrough that made high-speed, two-cycle engines possible (Figs. 1-8 and 1-10). The fresh charge enters the cylinder through multiple transfer ports. Exit ramps direct flow toward the

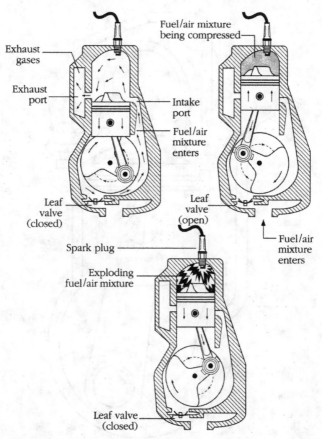

FIGURE 1-8. *Unlike third-port engines which mount their carburetors low on the cylinder barrel, reed-valve engines have their carburetors on the crankcase. The engine shown is cross-scavenged, as indicated by the deflector on the piston crown.*

chamber roof and away from the exhaust port. The charge streams merge to create a vortex that drives the exhaust gases out before it. But even at wide open throttle (wot), when loop scavenging is most efficient, some of the charge escapes through the exhaust port. At part throttle, the vortex loses integrity and a third or more of the fuel charge short-circuits to the atmosphere.

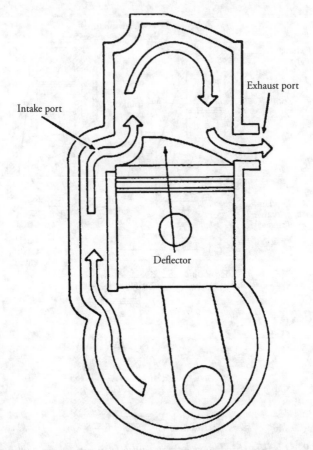

FIGURE 1-9. *Cross-scavenged engines offset the transfer port 180° from the exhaust port. The extrusion on the piston deflects the incoming charge up and away from the open exhaust port. Note that when assembling these engines that the steep side of the piston deflector faces the inlet port.*

Displacement

Displacement—the volume swept by the piston—is the basic measurement of engine size as square footage is to houses, tonnage is to ships, or caliber is to firearms. All things being equal, an engine with twice the displacement of another should be capable of twice as much work.

Bore × bore × stroke × number of cylinders × 0.7858 = displacement

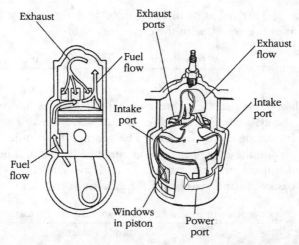

FIGURE 1-10. *Loop scavenging, pioneered in the 1930s by the German motorcycle manufacturer DKW and subsequently copied by everyone, was the major breakthrough in two-cycle engine development.*

If bore and stroke are expressed in inches, the formula gives displacement in cubic inches (cid). The Briggs & Stratton 96900 has a 2.56-in. bore and 1.75-in. stroke:

$$2.56 \times 2.56 \times 1.75 \times 1 \times 0.7858 = 9.01 \text{ cid}$$

European and Asian manufacturers express bore and stroke in millimeters (mm) and displacement in cubic centimeters (cc or cm^3). The Tanaka TBC 2501 two-stroke has a 34-mm bore and a 27-mm stroke:

$$34 \times 34 \times 27 \times 1 \times 0.7858 = 24{,}526.39 \text{ mm}^3 \text{ (cubic millimeters)}$$

Divide by 1000 to convert cubic millimeters to cubic centimeters:

$$24{,}526.39/1000 = 24.5 \text{ cm}^3$$

To express cubic centimeters in cubic inches ($in.^3$), divide by 0.061. The Tanaka in this example displaces 1.49 $in.^3$ or, rounded off, as is the custom, 1.50 $in.^3$. To work the other way multiply the cid by 16.387 to arrive at cubic centimeters. The 9.01 cid Briggs displaces 147.6 cm^3.

Compression ratio

The compression ratio (CR) is the measure of the "squeeze" put on the air/fuel mixture prior to ignition. Four-cycle-engine makers use the geometric compression ratio arrived at by dividing the cylinder volume at bdc, or the lower limit of piston travel, with the clearance volume at tdc.

$$\text{CR geometric} = \frac{\text{swept volume after bdc} + \text{clearance volume}}{\text{clearance volume}}$$

This formula assumes that the entire length of the compression stroke builds pressure in the cylinder. Actually the exhaust valve remains open for 20 crankshaft degrees or more after bdc. The geometric formula inflates the numbers, enabling these engines to be credited with CRs of 8 or 9:1. Two-stroke manufacturers are more scrupulous and calculate effective compression on the basis of piston travel after the exhaust port closes.

$$\text{CR effective} = \frac{\text{swept volume, after exhaust port closed} + \text{clearance volume}}{\text{clearance volume}}$$

Up to a point, increasing the CR results in more power or, if power is not a priority, better fuel economy. The limit is imposed by the tendency of the fuel to detonate. Normal combustion proceeds in an orderly manner with the flame front moving outward from the spark plug to occupy the entire combustion chamber. Detonation occurs when islands of unburned fuel, heated and compressed by the expanding flame front, spontaneously explode. Were air-cooled engines not so noisy, detonation would make itself heard as the metallic click old-timers called "spark knock." Persistent detonation flattens the rod bearings and blasts holes in the piston crown.

Horsepower, torque, and rpm

James Watt (1736–1819) needed a metaphor to convince mine owners of the worth of his pumping engines. They used animal power—horses condemned to a turnstile—to work the pumps that kept the mines dry. Combining observation with guesswork, Watt announced that the average mine pony could lift 550 lb per second or 33,000 lb per minute. That was one horsepower, an almost magical term that bridged the gap between animal and mechanical power. Watt's engines worked as fast as 200 horses and needed neither stabling nor fodder. Of course, the engines needed coal, but that was no problem for coal miners.

Horsepower is the rate of performing work and depends upon rpm and the twisting force, or torque, applied to the crankshaft. One lb/ft of torque is the instantaneous force exerted on a 1-ft-long bar by a weight of one pound (lb). Ph.D. dissertations have been written about the subtleties of the relationship between torque and horsepower, but for us it is enough to say that a powerful lawnmower races through high grass. Torque appears as the stubbornness that keeps the mower from stalling in the tough spots. For vehicles, horsepower translates as speed and torque as acceleration or as the ability to lug in high gear without downshifting. The flatter the torque curve relative to rpm, the less the engine bogs under load.

To determine an engine's torque, researchers mount it on a dynamometer and measure the braking force needed to bring the engine to a stop. The braking force expressed in lb/ft (or lb-ft—the terms mean the same) is the torque developed at the test rpm. Once this is known, brake horsepower (bhp) is derived by this formula:

$$bhp = \frac{torque \times 2\pi \times rpm}{33,000}$$

Because 2 pi (2π) divided by 33,000 is the same as 1 divided by 5252, the equation can be rewritten as:

$$bhp = \frac{torque \times rpm}{5252}$$

An engine that produces 2 lb/ft of torque at 2600 rpm develops:

$$bhp = \frac{2 \times 2600}{5252}$$
$$= 0.990$$

which would be rounded up to 1.0 bhp.

As with all measurements, we vacillate between English and metric units for horsepower and torque. Table 1-1 lists how to convert these terms to the more familiar English units.

Figure 1-11 illustrates a typical set of performance curves. Maximum torque occurs at roughly two-thirds of the rpm that provides maximum power for this and other small engines. As rpm increases, so does power output up to the point that friction and pumping loses intervene.

The devil is in the details or, in this case, the conditions under which torque is measured and horsepower is calculated. Briggs and most other manufacturers determine torque according to the Society of Automotive Engineers (SAE) code J1940 protocol and calculate horsepower by SAE J1995. In-house technicians break in the engines, remove carbon accumulations, and adjust the timing and fuel delivery with laboratory precision. Dyno tests are made without mufflers, air cleaners, and other power-robbing accessories at an ambient temperature of 76°F and at an elevation of 327 ft above sea level.

TABLE 1-1. Horsepower and torque

Metric nomenclature	To convert to U.S. nomenclature
metric hp (PS, CV, or DIN)	PS to hp: divide PS by 1.01387
watts (W)	W to hp: divide W by 746
kilowatts (kW)	kW to hp: divide kW by 0.746
Torque	
newton-meters (N-m)	N-m to lb-ft: divide N-m by 0.7373
kilograms force meter (kgf-m)	kgf-m to lb-ft: multiply kgf-m by 7.2330

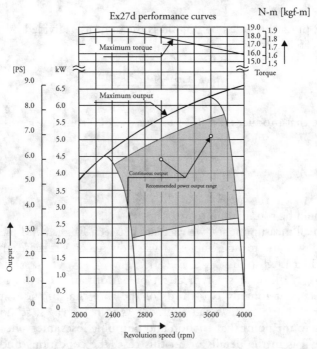

FIGURE 1-11. *Subaru performance curves differ from most in that engines are tested with the muffler and air cleaner installed. Note that the maximum power output, or PS, works the engine harder than the manufacturer recommends. The continuous power output is even less. Many small engines have a built-in power limiter, in that the governor caps rpm at a figure below its rated power output.*

The manual used by Briggs applications engineers warns that actual, in-service horsepower will be 80 percent of advertised. While small-engine horses may be ponies, the numbers have significance for comparative purposes. A Kohler 7.0-hp engine should be able to do as much work as a Subaru or Honda engine with the same hp rating.

However, some customers feel cheated and have gone to court with charges of false advertising. This may be why Briggs dropped horsepower ratings for its smaller engines and why Kawasaki adopted the SAE J2723 test protocol. According to Kawasaki, this strict rating method ensures that engines deliver at least 98 percent of rated horsepower. Third parties conduct the dynamometer tests. Horsepower numbers derived by this standard are preceded with "SAE."

Emissions

Edmunds Incorporated, a company devoted to automotive technology, compared the exhaust emissions of two leaf blowers against a 6200-lb Ford Raptor crew cab pickup truck. Three classes of emissions were of interest:

- Nonmethane hydrocarbons (NMHC), some of which are known carcinogens,
- Oxides of nitrogen (NO_x) that react with HC and other compounds to form acid rain and smog,
- Carbon monoxide (CO), a lethal gas that displaces oxygen in the bloodstream.

The 411-hp truck was strapped to a chassis dynamometer and run through the standard EPA emissions-test cycle. The leaf blowers—a 50.8-cc two-stroke Echo PB-500T backpack model and a four-stroke 30-cc Ryobi handheld unit—were run in their normal modes, that is, at wide open throttle interspersed with periods at idle.

The four-stroke Ryobi emitted more than six times the NO_x and NMHC per horsepower-hour than the Raptor. The two-stroke Echo did much worse, spewing out more than 23 times as much CO and 299 times more NMHC than the behemoth truck. Edmunds calculated that one would have to drive a Raptor from North Texas to Alaska, a distance of 3887 miles, to equal the emissions of a half-hour's work with the two-stroke leaf blower.

Poor scavenging accounts for the Echo's high levels of NMHC and why this and other small two-strokes generate comparatively little NO_x. Some exhaust gases remain in the cylinder after scavenging to quench the flame below the temperature that disassociates atmospheric air into nitrogen and oxygen. CO readings are a function of the amount of fuel burned. The Echo used nearly twice as much fuel per horsepower-hour as the four-cycle Ryobi.

Homeowners who use their garden equipment on weekends may be able to weather this sort of exposure without obvious health effects. The prognosis is different for groundskeepers working behind these machines eight or ten hours a day, especially when we factor in the risks of silicosis from breathing dust generated by leaf blowers.

Regulations

Health and environmental concerns about small engines took on force in 1995 when the California Air Research Board (CARB or ARB) set exhaust emissions limits for engines sold in that state. The federal Environmental Protection Agency then imposed national limits on the CARB model. Since then, both agencies have moved to further limit exhaust emissions (Table 1-2). The EPA estimates that the 100 million small engines in the United States account for 5 percent of the total air pollution, with two-stroke engines responsible for most of the contamination.

Limiting exhaust emissions involves a tradeoff between HC and NO_x. Increasing the amount of air in the air/fuel mix, that is, leaning the mixture,

TABLE 1-2. **Current small-engine emissions limits**

CARB Tier 3 emissions limits adopted in 2004

Displacement	HC + NO_x				CO
	2005	2006	2007	2008+	
<50 cc	50	50	50	50	536
50–80 cc	72	72	72	72	536
80–225 cc	16.1	16.1	10.0	10.0	549
>225 cc	12.1	12.1	12.1	8.0	549

EPA Phase 3 emissions limits fully in force for model year 2012

Non-handheld engines	HC + NO_x	CO
Class I (<250 cc)	10	610
Class II (250 cc and larger)	8	610
Handheld engines		
Class III (<20 cc)	50	805
Class IV (20 cc to <50 cc)	50	805
Class V (50 cc and larger)	72	603

Emissions limits are given in g/kW-h, and 1 hp = 746 W.

raises the flame temperature so that less HC escapes combustion. But these same high flame temperatures promote NO_x by disassociating atmospheric air into nitrogen and oxygen, which then recombine into various noxious compounds. The EPA and CARB acknowledge the quandary by combining the two classes of emissions into a single number.

The first round of regulatory limits was met by tweaking air/fuel mixtures and making other simple modifications. But CARB Tier 2 standards and EPA Phase 2 standards, phased in between 2001 and 2007, required more serious responses, especially for the two-strokes that CARB's Richard Varenchik described "as among the dirtiest engines on the face of the earth." Were these engines to survive, they had to meet emissions limits while remaining in character, that is, simple, lightweight, and aggressively powerful.

Stratified charging

RedMax Komatsu's Strat Charged engine, introduced in 1998, may well be the most significant advance in two-stroke engines since loop scavenging. The RedMax "airhead" met CARB Tier 2 emissions requirements with room to spare and, as an added benefit, cut fuel consumption by 30 percent. At first customers were reluctant, assuming that the soft exhaust note meant reduced power. Acceptance came when Stihl purchased 60,000 Strat Charged engines to make its products legal in California. Echo, Stihl, Tanaka, and other manufacturers followed up with their own versions of the technology.

Figures 1-12 through 1-14 illustrate how the RedMax works. The layout is similar to conventional two-strokes, except that the transfer ports—the ports that admit fuel to the cylinder from the crankcase—connect via reed valves to a pair of air intake tubes. In the drawings the transfer ports have been renamed "scavenger ports" in recognition of their exhaust-purging function. And, unlike conventional carburetors, the Walbro WYA has a second, throttle-controlled barrel for fresh-air delivery.

Hybrids

String trimmers, brush cutters, chainsaws, and similar handheld tools must be able to operate at any angle, even upside down. When a conventional four-stroke engine is inverted, crankcase oil migrates to the overhead valve chamber and into the carburetor through the crankcase ventilation tube. Should the engine continue to run, the low or nonexistent crankcase oil level results in rapid failure. Hybrid engines combine four-stroke operation with special provisions to isolate the oiling system from the effects of gravity.

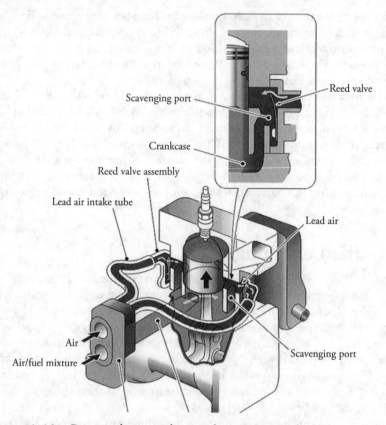

FIGURE 1-12. *During the upstroke, crankcase vacuum draws a mixture of air and fuel into the crankcase in the manner of conventional two-strokes. Crankcase vacuum also opens the reed valves to admit fresh air to the lead air intake tubes.* RedMax Komatsu Zenoah America, Husqvarna Division

Because exhaust and intake events are each allocated a piston stroke, hybrids easily meet current emissions limits. These engines start easier than two-strokes, run more quietly, and cut fuel costs by half. On the other hand, hybrids are more expensive, heavier, and less powerful than the engines they replace. The flatter torque curve can compensate for the lack of full-throttle power, but the extra weight is felt when using the tool.

The Stihl 4-Mix (Figs. 1-15 through 1-18) and Shindaiwa Hybrid 4 engines lubricate with a 50:1 two-stroke premix, made possible by replacing the oil-hungry plain bearings in conventional four-strokes with needle and ball bearings. Of course, the oil-impregnated fuel generates HC emissions, but the inherent cleanliness of four-stroke engines more than compensates.

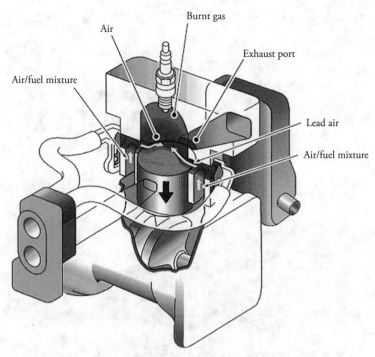

FIGURE 1-13. *On the downstroke, the piston pressurizes both the air/fuel mixture in the crankcase and air in the air intake tubes upstream of the reeds. The reed valves are closed to isolate the carburetor from lead air backflow. As the piston continues to fall, it cracks open the scavenging ports to admit air earlier into the cylinder than the air/fuel mixture.*
RedMax Komatsu Zenoah America, Husqvarna Division

Commercial users appreciate that there is no oil level for their workers to check.

Oil mist lubrication, developed in the 1930s and used widely in the machine tool industry, was put into series production for small engines by Ryobi in response to Phase 2 EPA regulations. The crankshaft counterweight functions as an atomizer and generates enough pumping force to move the oil mist through the engine. It sounds simple, but the company spent a reported $10 million on the project. Somewhat similar approaches were used by Husqvarna and Briggs.

Honda's Mini 4-Stroke series currently includes four commercial-grade engines with displacements ranging between 25 and 49 cc (Fig. 1-19).

Honda's lubricating system is, in a word, elegant (Fig. 1-20). A lightweight plastic oil reservoir just inboard of the rewind starter houses a two-bladed

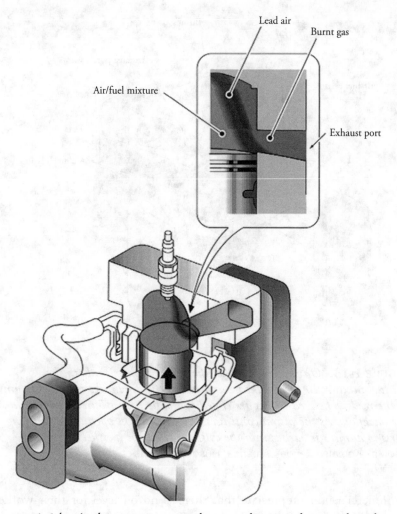

FIGURE 1-14. *As the piston rises on the upstroke, it pushes out the exhaust gases through a cushion of lead air, most of which short circuits. Only a fraction of the air/fuel mixture escapes out the exhaust.* RedMax Komatsu Zenoah America, Husqvarna Division

rotor, or slinger (56). The slinger atomizes the lube oil into a fine mist that provides sufficient lubrication and, unlike liquids, moves under slight pressure differentials. Energy for the system is created by piston movement.

During the upstroke, atomized oil moves into the crankcase through a crankshaft drilling that, regardless of the attitude of the engine, remains above the liquid level in the reservoir. The oil mist lubricates the main

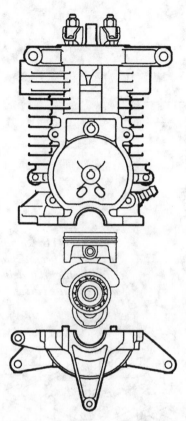

FIGURE 1-15. *Stihl's horizontally split crankcase eliminates the hassle of detaching main bearings shrunk into full-circle housings.*

bearings, crankpin, and piston. On the downstroke the reed valve (61) opens to relieve crankcase pressure. The vented pressure pulses, consisting of oil mist and blowby gases, pass into the hollow camshaft (21). The cam functions as a centrifugal oil/gas separator. Oil droplets flung out of the cam lubricate and cool the exposed valve gear, while blowby gases vent to the air cleaner (4) for recirculation. The oil then returns to the reservoir through a pipe (78) where it condenses back into a liquid.

By opening in response to pressure pulses, the reed valve creates a partial vacuum in the crankcase and less of a depression (due to the restriction imposed by the crankshaft drilling) in the oil reservoir. Thus the lowest pressure is in the crankcase; a somewhat higher, but less-than-atmospheric, pressure is in the oil reservoir; and atmospheric pressure punctuated by oil-laden pressure pulses is in the valve chamber.

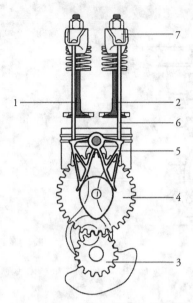

FIGURE 1-16. *Sophisticated mathematics must have been deployed to arrive at a cam profile that drives both the intake and exhaust valves.*

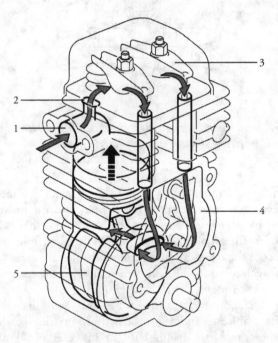

FIGURE 1-17. *As the piston climbs on the compression stroke, it creates a depression in the crankcase. The crankcase vacuum passing through a cutout on the piston skirt depressurizes the overhead-valve chamber. Oil-impregnated fuel enters through inlet pipe 1; moves into the valve chamber through access port 2; and, drawn by the vacuum, collects in the crankcase.*

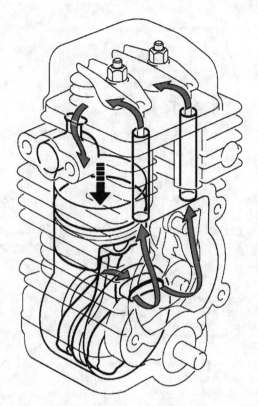

FIGURE 1-18. *During the inlet stroke the falling piston pressurizes the fuel charge in the crankcase, an action that forces the fuel charge back up to the valve chamber and into the cylinder through the open intake valve.*

The pressure differential between the crankcase and reservoir draws oil into the crankcase, which is then carried to the valve chamber by pressure pulses, and back to the low-pressure reservoir.

Fuel injection

As described in Chap. 4, electronic fuel injection (EFI) can reduce HC emissions by 40 percent and provide dramatic gains in fuel economy. Currently applications are limited to upscale Kawasaki, Kohler, Briggs, and Subaru four-cycle engines, but the real breakthrough will come if this technology can be adapted to two-stroke handheld engines, as Piaggio has done for its Vespa scooters.

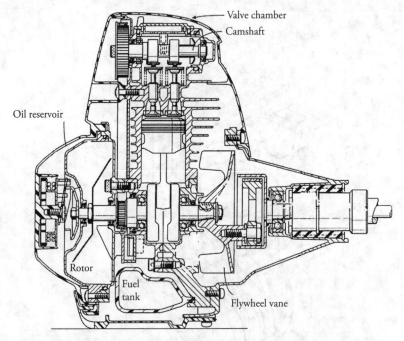

FIGURE 1-19. *The second-generation ohc Mini 4-Stroke was designed with a low polar moment of inertia. The short crankshaft, compact ohc valve gear, and underslung fuel tank concentrate mass around the center of gravity to improve the responsiveness of handheld tools.*

Catalytic converters

Stratified-charge two-stroke engines meet current U.S. and European emissions limits, but these limits will get tighter in the future. Aside from long shots like direct in-cylinder injection, the catalytic converter offers the best hope for cleaning up two-stroke emissions. Probably as preparation, Stihl, Husqvarna, its group partner Jonsered, Dolmar, and Echo offer these devices on professional chainsaws and, more rarely, on blowers and other handheld equipment.

It's instructive to see how the pioneering users of catalytic converters adopted this expensive and fragile technology to small engines. Cat converters are not a bolt-on proposition. The Carb Tier 2 Tanaka functions like an ordinary, crankcase-scavenged two-stroke except that the leads to the transfer passages between the crankcase and cylinder are narrow. This restriction

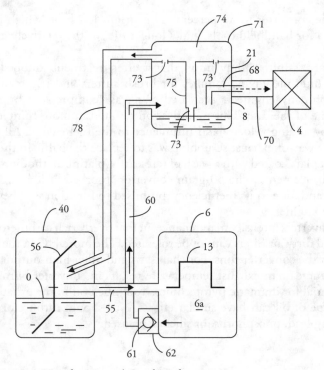

FIGURE 1-20. *Honda Mini 4-Stroke oiling circuit.*

improves combustion by adding velocity and heat to the fuel charge. HC and CO that survive then pass into a catalytic converter where

$$HC + O_2 \rightarrow H_2O + CO_2$$

The heat generated by this exothermic reaction raises exhaust temperature to 1000° and enables carbon monoxide to combine with oxygen to form carbon dioxide:

$$CO + 1/2\ O_2 \rightarrow CO_2$$

Other reactions involving CO release water and hydrogen.

The converter in the first version of Tanaka's PureFire technology scrubbed 40 percent of the HC and CO in the exhaust. Thanks to a redesigned combustion chamber, the second-series PureFire reduces the role of the converter to 30 percent. This means less heat, which is always a problem for handheld tools, and a longer life for the relatively expensive converter.

Husqvarna's initial foray into exhaust after-treatment, dubbed E-tech, involved multiple engine families for brush cutters and trimmer applications. Company engineers increased crankcase compression by reducing the volume of the case. This modification resulted in more complete combustion by adding velocity and turbulence to the fuel charge. All that was needed by way of exhaust scrubbing was to replace the baffle in the muffler with a flat plate coated with a wash of rare-earth (platinum, rhodium, and/or palladium) elements. The 30-gram converter absorbed little heat from the reaction, and, due to its inefficiency, promised to function well beyond the life of the engine.

Cat converters intensify maintenance, in that overly rich carburetor adjustments and dirty air filters cannot be tolerated. The more HC in the exhaust, the hotter the converter runs. For chainsaws and other high-output engines, a 300°C increase in exhaust temperature results in converter burnout with the potential for dramatic pyrotechnics. Too much oil in the exhaust or the wrong type of oil can have similar effects. Nor should carburetor cleaners containing lead, phosphorus, or silicone be used.

Fuel

Tetraethyl lead had been used as an octane booster for 60 years when it was outlawed in 1986. The Environmental Protection Agency (EPA) was moved to act because lead poisons the catalytic converters (and people, but that was a secondary consideration). Leaded gasoline remains available for piston-engined aircraft and as motor fuel in some developing countries.

Denied lead, most refiners turned to methyl tertiary butyl ether (MTBE), which boosts octane by adding oxygen to the fuel. Unfortunately MTBE contaminates groundwater, making it undrinkable. California outlawed the additive in 2003, and most other states followed suit. Today nearly all gasoline sold in this country is oxygenated with ethanol alcohol. E10 gasoline contains 10 percent ethanol by volume, E15 contains 15 percent, E30 contains 30 percent, and so on. E85 has the highest ethanol content, ranging between 51 and 83 percent depending upon the season and locality. The blend is labeled on the gas station pump in most states.

Out of necessity, manufacturers have adapted small engines to E10. Changes to the composition of fuel lines, gaskets, and plastics used in fuel systems solved or, at least mitigated, the immediate problems. But the corrosion associated with long-term exposure to ethanol has yet to be addressed. Ethanol is hydroscopic, which means that it absorbs water from

FIGURE 1-21. *What water does to carburetors.* Photo by Tony Shelby

the atmosphere. The acids produced by a water/alcohol mix are death on pot-metal carburetor castings (Fig. 1-21).

Because of its affinity for water, E10 should be used within a few weeks of purchase. Fuel stabilizers—products like Sea Foam or Sta-Bil—extend storage life, but do not eliminate all problems. The better option is to avoid ethanol altogether. Some (mostly rural) gas stations, marinas, and airports catering to private aviation carry ethanol-free gasoline. Similar gasolines, some with a 50:1 oil mix for two-stroke engines, are sold by the quart at Lowe's and other big-box stores.

Gas cans

CARB and the EPA mandate that new gasoline cans have childproof locks, spring-loaded lids, no vent ports, and spouts that make a vapor-proof seal with engine fuel tanks. But simple, everyday objects like shovels or gas cans are difficult to improve (Fig. 1-22). Several of these new containers have collapsible spouts that wet your hands when you extend them, others must be held at an exact angle to pour, and the absence of a vent means that flow proceeds in gulps and ends in spills.

The Occupational Safety and Health Administration (OSHA) has different rules for gas cans used on worksites. These cans are much more convenient to use and, because they come with a fire screen, appear to be

FIGURE 1-22. *Girl with a gasoline can in Puerto Rico, circa 1938.*
U.S. Library of Congress

safer than the EPA versions. Not being a lawyer, I do not know if it is legal to use OSHA-approved gas cans at home.

The best engineered gas cans were used by the German army during World War II (Fig. 1-23). The cans were copied by the British, who called them "jerry cans," and they continue to be issued to NATO forces. Features include heavy-gauge steel construction, continuously welded seams, internal vents, and cam-operated caps with safety pins to prevent accidental opening. At this writing, Wavian brand NATO cans, available from Amazon and other outlets, are the only ones to carry CARB, EPA, and U.S. Department of Transportation approval. These cans, which also comply with German, Swedish, and Australian standards, come in 5-, 10-, and 20-liter (5.28-gallon) capacities.

Durability

Table 1-3 summarizes what the EPA has learned about how long small engines live. The findings are based on manufacturer's data, surveys of owners, and information collected by CARB.

FIGURE 1-23. *Wavian NATO gas cans are expensive, but durable and legal. The 20-liter container shown retails for about $80, and the 5-liter version is a few dollars cheaper.*

If we exclude throwaway products, homeowners can expect to pay $150 for a string trimmer that has a median life expectancy under full load of 35.3 hours. The median life expectancy is the number of hours that half of the machines remain repairable. Professional-quality string trimmers retail for about $350 and have eight times the median life expectancy. A respectable homeowner-quality lawnmower costs about $400 and has a 50 percent chance of dying within 48 full-load hours. Paying four times as much for a professional machine increases median life by almost six times. The same pattern holds for other garden and landscaping equipment: money spent beyond the basic cost of the machine provides a disproportionate benefit in durability.

All 2010 and newer engines carry a sticker that, among other useful information, shows for how many operating hours the engine will remain in compliance with emissions regulations (Fig. 1-24). The emissions compliance period is the same as the effective engine life as determined by engineering studies, field experience, and warranty claims and other data the EPA and the manufacturer think pertinent.

TABLE 1-3. Median life of 25-hp and smaller nonroad engines

Type	Use	HP (min.)	HP (max.)	Median life (hours at full load)	Annual hours	Load factor	Median life (years)
Lawnmowers	Residential	1	6	47.9	25	0.33	5.8
	Commercial	1	6	268.0	406	0.33	2.0
String trimmers, edgers	Residential	0	3	35.3	9	0.91	4.3
	Commercial	0	3	286.8	137	0.91	2.3
Chainsaws	Residential	0	6	39.2	13	0.70	4.3
	Commercial	0	6	191.0	303	0.70	0.9
Leaf blowers, vacuums	Residential	0	6	40.4	10	0.94	4.3
	Commercial	0	6	609.7	282	0.94	2.3
Snow blowers	Residential	0	6	12.3	8	0.35	4.4
	Commercial	0	6	209.4	136	0.35	4.4
Rear-engine riding mower	Residential	3	16	79.3	36	0.38	5.8
	Commercial	6	16	627.0	569	0.38	2.9
Lawn and garden tractor	Residential	3	25	114.8	45	0.44	5.8
	Commercial	6	25	920.0	721	0.44	2.9

Source: EPA, "Median Life, Annual Activity, and Load Factor Values for Nonroad Engine Emissions Modeling," Report No. NR-005d, July 2010.

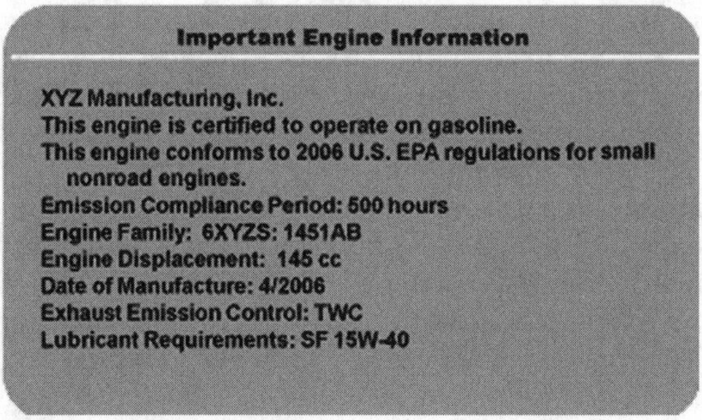

Important Engine Information

XYZ Manufacturing, Inc.
This engine is certified to operate on gasoline.
This engine conforms to 2006 U.S. EPA regulations for small
    nonroad engines.
Emission Compliance Period: 500 hours
Engine Family: 6XYZS: 1451AB
Engine Displacement: 145 cc
Date of Manufacture: 4/2006
Exhaust Emission Control: TWC
Lubricant Requirements: SF 15W-40

FIGURE 1-24. *EPA compliance sticker.*

Compliance periods are expressed by a letter grade:

Engines displacing more than 80 cc and less than 225 cc:

> A = 500 hours
> B = 250 hours
> C = 50 hours

225-cc and larger engines:

> A = 1000 hours
> B = 500 hours
> C = 250 hours

Cheap, throwaway engines have compliance periods of 50 hours. Midrange engines should be good for 125 hours or so, and top-of-the-line products for 350 or more hours. These figures have statistical validity: if the sticker says 125 hours, the engine may die sooner, but odds are it won't live much longer than 125 hours, no matter how often you change the oil or clean the air filter. It's interesting that while manufacturers can choose compliance periods of as long as 1000 hours, none to my knowledge have done so.

The duration of the factory warranty is another indication of the confidence manufacturers have in their products. Subaru provides a five-year across-the-board warranty for its EH and EX four-cycle engines. Tanaka warranties its two-stroke equipment for seven years residential, two years commercial, and one year rental. With some exceptions, Honda covers its GX commercial-grade engines for three years regardless of the type of service. Honda GC and GCV engines are covered for two years in residential service and for three months in commercial, rental, or institutional service. Depending upon the importer, Chinese clones may come without any real warranty, or any warranty that will be honored short of argument. Note that factory warranties are in addition to the EPA-mandated two-year warranty for emissions-related components. In theory, the EPA and CARB emissions warranties provide free replacement for sacrificial items such as spark plugs and fuel filters, but don't count on it.

Rental agencies are, of necessity, experts in small engines. You can be sure that the brands they purchase are durable and have dealer support in the area. Professional groundskeepers are also a source of reliable information.

Weight is a rough indicator of survivability for four-stroke engines, and cast-iron blocks are widely considered the gold standard. Briggs stopped making iron engines in the mid-1960s in the United States, although the company still builds them for the domestic market in China. Cast-iron cylinder sleeves, or liners, extend the lives of the better aluminum-block engines from all manufacturers (Fig. 1-25).

While all modern two-strokes use ball and needle bearings to reduce lubrication requirements, only the best four-cycle engines support their crankshafts on ball bearings. Less-durable engines run their crankshafts directly against the aluminum block metal. Although worn aluminum cylinders can be machined oversize, replacing the engine is usually the better option.

FIGURE 1-25. *This Chinese clone is sleeved, but the sleeve is too thin to permit re-boring. The makers didn't bother to remove the sand casting marks on the piston.* Photo by Tony Shelby

FIGURE 1-26. *A half-crank engine.* Photo by Tony Shelby

Throwaway two-strokes have their crankshafts cantilevered off one or two main bearings on the output end of the shaft (Fig. 1-26). These half-crank engines can be recognized by the rewind starter mounted low, on the underside of the block adjacent to the centrifugal clutch. Rather than being machined out of bar stock, half-cranks often have counterweights pieced together from sheet-steel laminations. Welded-up steel stampings serve as connecting rods. Henry Ford had an impolite term for stamped steel: he called it "s__iron."

The definitive sign of two-stroke excellence is the use of nickel-silicon composite cylinder bore coatings rather than chrome or nickel. Developed by MAHLE and U.S. Chrome, these coatings consist of tiny particles of silicon carbide dissolved in a nickel matrix. As the engine runs, the piston rings scrape away the nickel and bear directly on the diamond-hard silicon. Cylinder wear becomes almost a nonissue.

2

Troubleshooting

Some mechanics can look at an engine, make a few tests, and tell what's wrong with it. Other mechanics, like gamblers down on their luck, throw parts at the problem. The structured approach outlined in this chapter takes most of the uncertainty out of the process.

Begin by replacing the spark plug(s). The owner's manual lists the recommended spark-plug brand and type number. Do not be misled by appearances: the original spark plug may appear clean and function perfectly outside of the cylinder but fail to fire under compression. Set the gap on the replacement plugs with a feeler gauge or ramp-type tool (Fig. 2-1). The specification is usually 0.7 or 0.8 mm (0.028 or 0.031 in.). Run the plugs in as far as you can by hand, and then using a wrench tighten 8- and 10-mm diameter plugs another quarter turn and 12-mm plugs three-eighths of a turn.

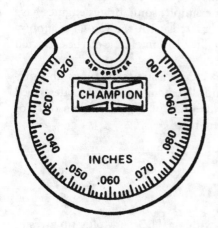

FIGURE 2-1. *A ramp-type gauge quickly and accurately sets spark-plug gaps. If you use a flat-bladed feeler, bracket the readings. For example, when the specification is 0.030 in., adjust the gap for the 0.029-in. blade to slip in easily and the 0.031-in. blade to make rubbing contact with the electrodes.*

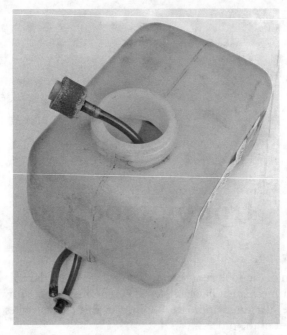

FIGURE 2-2. *A wire bent into a hook is used to retrieve in-tank filters. Note that these filters are weighted to sink below the liquid level in the tank. Omitting the filter enables the fuel line to rise to the surface and ingest air that leans the mixture and quickly destroys the engine.*
Photo by Tony Shelby

E10 gasoline has a shelf life measured in weeks. If there is any doubt about the fuel quality, drain the tank and add newly purchased gasoline from a clean container. Service the air-filter element and replace the fuel filter or filters with the correct factory part. For example, some Briggs engines have two-stage filtration with the primary filter rated at 150 microns (1 micron = 0.000039 in.) in series with a 75-micron filter. In-tank filters can be extracted with a coat hanger (Fig. 2-2).

Air filters richen the mixture as they clog. Plastic foam filters and pre-filters should be washed with water and detergent and oiled as described in Chap. 4. Pleated-paper filter elements must be replaced when dirty or wetted by gasoline from a flooded carburetor.

Basics

To run, an engine must have:

- Adequate compression.
- Fuel vapor in the right proportion with air. The spark-plug tip gives us an indication of cylinder fueling.

- A high-voltage spark for ignition.
- A free-flowing exhaust that permits the cylinder to evacuate.

No start

If the engine does not start after a half-dozen attempts, do not press the issue. Further cranking will flood the combustion chamber and make starting even more unlikely.

Compression

Compression can be measured on two- and four-stroke engines without automatic compression releases. Install a good-quality compression gauge (cheap gauges are inaccurate) in the spark-plug boss as described in Chap. 7. Switch the ignition Off, disengage the choke, and open the throttle wide. Use a tie wrap to hold spring-loaded throttle triggers open. Crank the engine as if attempting to start it. Repeat four or five times to obtain the highest gauge reading. Chainsaws and other high-performance two-strokes should develop 110–150+ psi cold cranking compression with the manual decompressor valve closed. Four-stroke engines without a compression release develop between 90 and 120 psi during cold cranking. Overhead-valve models generate higher compression than side-valve engines.

Most modern four-strokes have centrifugal compression releases that crack the intake or exhaust valve open during cranking. With these engines it is enough to sense the presence of compression as resistance on the starter cord or the rise and fall of the as sound made by the electric starter. Wear on the decompressor mechanism or excessive valve clearance defeats the compression release to produce cold-cranking pressures on the far side of 125 psi. The engine will kickback and snort, and the starter handle will recoil violently.

If you encounter low compression, squirt a little oil into the cylinder and spin the flywheel to distribute the oil over the bore and then retest. Should the gauge register a compression increase, the rings are not sealing. Removing the muffler on two-stroke engines will reveal if scoring is present on the exhaust-port side of the piston.

If oil has no effect on compression, the bore seal is good and the loss of compression is associated with the valves or head gasket.

Fuel

To determine if fuel is reaching the cylinder, close the choke (if present) or cycle the primer bulb to wet the carburetor throat with fuel. Crank the engine over five or six times. The spark-plug tip should be damp with fuel or at least smell of gasoline. When in doubt, spray a small amount of carburetor cleaner into the cylinder, replace the spark plug, and crank. If the engine runs a few seconds and dies, you can assume that ignition and compression function normally and that fuel starvation is the problem. For additional verification, remove the air-filter element and spray a small amount of carb cleaner into the carburetor air horn. Replace the filter element, open the throttle, and crank. A healthy, but otherwise fuel-starved engine will run a few seconds longer than in the previous test. An engine that starts with fuel in the chamber and does not start with fuel sprayed into the carburetor:

- Leaks air into the induction tract, as, for example, at the carburetor flange or the main-bearing seals on two-stroke engines
  or
- Has an intake valve or reed valve problem

If loosening the fuel line at the carburetor fitting demonstrates that fuel reaches the instrument and if the engine runs from fuel sprayed into the air horn, the fault lies in the carburetor. Diaphragms lose elasticity, and tiny fuel passages become obstructed with varnish and silt. Some four-stroke carburetors have an anti-dieseling solenoid at the fuel inlet or on the bottom of the float bowl. This solenoid, recognized by the attached electric wire, should click open when the ignition key is turned On.

Spark

Figure 2-3 illustrates two ways of hooking up a spark tester. Magnetos and capacitive-discharge ignitions (CDIs) generate thick, blue sparks capable of blistering paint. Briggs & Stratton Magnetron systems must be cranked at 350 rpm to produce a spindly orange spark that disappears in bright sunlight. But appearances aside, Magnetrons are among the most reliable solid-state ignition systems and are certainly the least expensive to replace.

If spark output is nonexistent or intermittent, electrically isolate the ignition module from the kill switch, low oil-level switch (Fig. 2-4), and other interlocks present on garden tractors and riding mowers. Any of these switches can malfunction to deny ignition and/or electric starting. Rotary lawnmowers can lose ignition if the blade hits something hard and the flywheel key—intended to be sacrificial on Briggs engines—shears or distorts. See Chap. 3 for additional information.

Crankshaft binds or freezes during cranking

Remove the spark plug(s) and turn the flywheel by hand. If the crankshaft moves through its full arc without protest, the problem may be with the

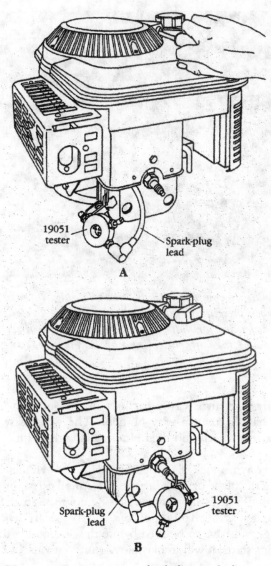

FIGURE 2-3. *Use an ignition tester to check for spark during cranking (A) and to detect misfires in a running engine (B).*

starter. A rewind starter will bind if out of alignment with the crankshaft hub. As described in Chap. 6, many things beginning with the battery can go wrong with electric starters. It is also possible that the equipment the engine drives is the source of the difficulty. Check for dragging clutches, overly tight belts, and frozen bearings on driven components.

FIGURE 2-4. *Utility-engine ignitions short to ground through the kill switch and, when present, an oil-level shut-off switch. In addition, riding mowers and garden tractors have safety interlocks on the driver's seat, transmission, power take off, clutch, and brake.* Photo by Tony Shelby

A crankshaft that turns easily for part of a revolution and then gets progressively harder to turn may be bent. The worst-case scenario is a crankshaft that freewheels and then binds solid with a clink. That means a thrown rod.

Lawnmowers hydro-lock if propped up with the cylinder head down as when servicing the blade or draining the oil. (An oil extractor pump such as the Oregon 88-405 saves a lot of trouble.) Remove the spark plug, turn the flywheel to displace as much oil as possible, and try to start the engine. Several dry spark plugs will have to be used before the oil-rich mixture can ignite. Once the engine starts, it will smoke like Mount Vesuvius, but that will pass.

Kickback during cranking

Engines with insufficient flywheel mass—a category that includes many handheld and gardening tools—hump and snap back as the starter cord is pulled. It is as if the engine is in revolt. A loose rotary mower blade, which contributes to lack of inertia, intensifies the unpleasantness and can make starting impossible. Ditto for an inoperative compression release.

Loss of power and/or bogging under acceleration

Do the routine things first—that is, replace the spark plug(s), fuel filter, and air-filter element, and verify that clutches and other driven components turn easily.

- Lean air/fuel mixtures are the most common reason for loss of power. Carburetors, and especially modern emissions-compliant carburetors, go lean as fuel passages clog and diaphragms stiffen. Air leaks at the carburetor mounting flange or at two-stroke crankshaft seals can also rob the engine of fuel. Insufficient fuel bleaches the spark-plug tip white and causes the engine to bog under load or when the throttle is rapidly snapped open by hand. If partially closing the choke restores performance, you can be confident that the lean air/fuel mixture diagnosis is correct. Chapter 3 has a section on reading spark plugs; Chap. 4 goes into detail about fuel system problems and describes how to test two-stroke crankcase integrity.
- Remove the valve cover on ohv and ohc engines. Some oil will spill as the cover comes free. Check the valve lash as described in Chap. 7.
- Two-stroke engines clog exhaust ports, mufflers, and spark arrestor screens (Figs. 2-5 and 2-6). Mufflers and arrestor screens can be cleaned by heating with a butane torch. To clean exhaust ports, remove the muffler, turn the flywheel to block the port with the piston skirt, and carefully scrape off the accumulated carbon.
- Overheating can contribute to loss of power. If you suspect that this is the problem, remove the shrouding and clean the cylinder fins as shown in Figure 2-7.

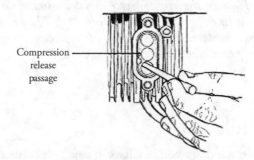

Compression release passage

FIGURE 2-5. *Two-stroke exhaust ports should be periodically cleaned. Some Tecumseh models have the compression bleed port shown in the drawing, which can clog and make starting difficult.*

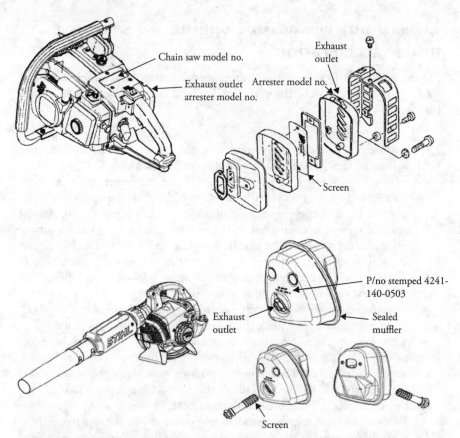

FIGURE 2-6. *Most spark arrestors, like the Echo/John Deere unit in the upper drawing, take the form of a rectangular screen sandwiched inside the muffler. The Stihl tubular arrestor in the lower drawing is serviced with the muffler in place. Heating the arrestor screen with a butane torch converts carbon deposits to a fine ash that can be brushed off. The process is messy and should be done outdoors while wearing eye protection.*
U.S. Department of Agriculture

Refusal to idle

Clean the carburetor as described in Chap. 4, paying special attention to the low-speed circuit. Note that small engines have a limited rpm range and few will "Cadillac" at 1000 rpm or less.

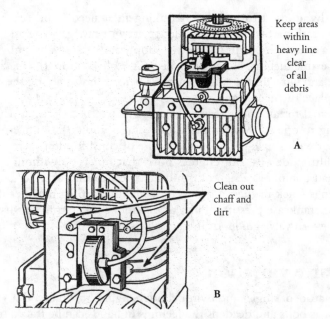

Keep areas
within
heavy line
clear
of all
debris

A

Clean out
chaff and
dirt

B

FIGURE 2-7. *Cooling fins collect dust and debris.* Briggs & Stratton Corp.

Engine runs a few minutes and quits

Connect a spark tester in series with the spark plug and start the engine. Watch the spark arc. If ignition failure causes the engine to stop running, the flywheel will coast to a stop without generating a spark. Replace the ignition module.

Fuel starvation, described under the "No start" heading earlier, is another possibility. Replace the fuel filters; clean the carburetor and, if present, the fuel pump; and replace all diaphragms. Also, verify that the carburetor solenoid valve functions and that the fuel tank vent—incorporated into the gas cap or remotely located and connected to the tank by a hose—is open. If running the engine with the gas cap loose solves the problem, the vent is clogged.

Exhaust smoke

Black smoke and loose, fluffy carbon deposits on the spark plug indicate that the engine is receiving more fuel than it can burn. Assuming that the air filter is clean and the choke opens fully, the most likely cause is an improperly adjusted carburetor. A leaking inlet needle and seat valve on float-type carburetors floods the inlet tract with raw gasoline and sets the stage for a serious fire. Shut off the fuel supply at the tank valve or by squeezing the flexible fuel line closed with Vise-Grip pliers, move the machine out of doors, and wipe up the spilled gasoline before disassembling the carburetor to replace the needle and seat.

It's normal for air-cooled four-stroke engines to use oil, even when new. According to one study, consumption averages about 0.5 ounce per operating hour with single-grade oils. The rate of consumption increases when using multi-grade oils. But engines burn that much oil without smoking. Blue exhaust smoke is a sign of trouble. Expect to find cylinder-bore wear, piston rings stuck in their grooves, or a malfunctioning breather that fails to relieve crankcase pressure. Puffs of oil smoke on startup can usually be traced to worn valve guides or intake valve seals.

Excessive vibration

It is the nature of single- and twin-cylinder engines to vibrate. But vibration that loosens bolts and deadens the feelings in hands can be traced to failure of the rubber insulation on portable tools or to a bent blade or crankshaft on rotary mowers.

It's worth repeating that working on the underside of a mower, tiller, shredder, or tractor without removing or disabling the spark plug(s) is dangerous. Do not trust the kill switch. Should the ignition system function, a slight movement of the crankshaft can start the engine.

To determine if mower blades are bent, mark a point on the deck adjacent to a blade tip. Rotate the flywheel 180° and verify that the other blade tip aligns with the mark. Placing the blade on a flat surface will make it easier to detect bends and twists. To determine if the crank is bent, focus on the bolt hole in the end of the crankshaft while a helper spins the engine over with the spark plug removed and the ignition grounded. If there is a perceptible wobble, a new crankshaft is in order.

3

Ignition and related systems

During the 1980s solid-state switching elements replaced the troublesome contact points in magnetos. Once that was done, it was relatively easy to include additional circuitry to retard the spark for easier starting and to advance the spark in concert with engine rpm.

Troubleshooting

Before faulting the ignition system,

- Replace the spark plug(s).
- Connect a spark tester between the high-tension lead and spark plug as shown back in Figure 2-3. If the system delivers a consistent spark, the ignition system can be considered good. If no spark or an erratic spark is present:
  - Inspect the flywheel key for damage and replace as necessary.
  - Disconnect the ground lead to the ignition module. The ground lead goes to the kill switch on all engines and to various safety interlocks on garden tractors and riding mowers. Test for spark output. If spark is present, inspect the ground lead for shorts and test the stop switches.

Spark plugs

Replace or clean the spark plug(s) every 100 hours of operation or at the first sign of hard starting. As the spark plug ages, carbon accumulates on the insulator tip where it shunts ignition voltage to ground. Wire-brushing the carbon off the insulator tip usually is enough to get a fouled spark plug to fire. Northern Tool sells a small abrasive blaster for cleaning spark plugs. The tool uses an unspecified "non-silicon" abrasive and requires 90 psi of air pressure. Briggs & Stratton and several other small-engine makers warn that abrasives used to clean spark plugs inevitably find their way into the cylinder, but industrial users routinely clean plugs in this manner. Liquid cleaners should not be used because they compromise the dielectric strength of the ceramic insulator.

Remove congealed grease from areas where the spark plug mates against the cylinder head. Use of anti-seize compounds on the threads encourages overtightening. Silicone spray has less lubricity and offers some protection against the electrolytic corrosion that occurs when dissimilar metals are put into intimate contact. Then, using a wrench tighten the spark plug as described in Chap. 2. Or better, torque to specification:

Thread diameter	Torque limits
Flat-seated spark plugs w/gasket, aluminum head	
8 mm	5.8–7.2 lb/ft
10 mm	7.2–8.7 lb/ft
12 mm	10.8–14.5 lb/ft
14 mm	18.0–21.6 lb/ft
Conical-seated spark plugs w/o gasket, aluminum head	
14 mm	7.2–14.5 lb/ft
18 mm	14.5–21.6 lb/ft

Heat range

Heat range is a measure of how quickly the spark plug disposes of combustion heat. The deeply tapered noses on hot plug insulators absorb and hold combustion heat, while cold plug insulators have less exposure and less heat retention (Fig. 3-1). When properly matched to the engine, a spark plug, however "hot" or "cold," maintains a tip temperature in the range of 500–850°C under varying loads (Fig. 3-2) and operating conditions (Fig. 3-3).

Reading spark plugs

With a used, but serviceable, spark plug of the recommended type installed, run the engine for 15 minutes or so under normal load. (A new spark plug may have to be run longer to develop color.) Cut the ignition without returning to idle. The insulator tip should be stained tan or russet brown. Excessively rich mixtures darken the tip with fluffy carbon or, in extreme cases, with unburned fuel (Fig. 3-4). High tip temperatures turn the insulator bone white, erode the electrodes, and may leave blue temper marks on the surrounding metal with disastrous effects on the piston (Figs. 3-5 and 3-6).

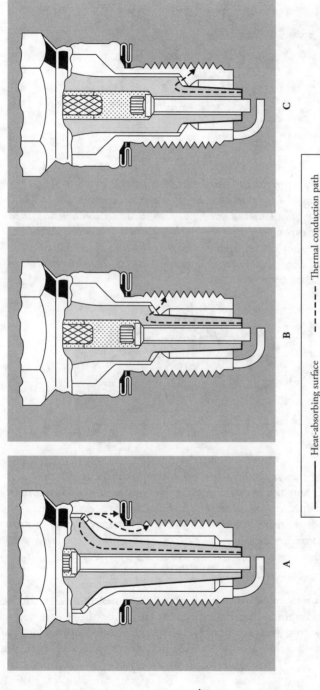

FIGURE 3-1. *The heat range depends primarily upon the length of the thermal path from the insulator tip to the relatively cool cylinder head. A hot plug (A) exposes a large area of the insulator to combustion heat that must travel high up in the spark-plug barrel to find release. A mid-range spark plug (B) has a shorter thermal path and a cold plug (C) has an even more direct thermal path.*

Heat-absorbing surface ----- Thermal conduction path

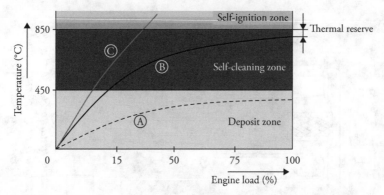

FIGURE 3-2. *The heat resulting from engine load affects spark-plug selection. Spark plug A is so cold that its tip temperature remains in the deposit zone. Spark plug C is too hot for the application. Under sustained load the tip would incandesce and ignite the mixture while the piston is deep in the compression stroke. Spark plug B is the best choice. It runs hot enough to burn off carbon deposits, but not so hot that it initiates pre-ignition.* www.briskusa.com

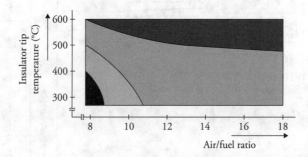

FIGURE 3-3. *The coolest temperature zone, shown on the lower right, occurs during cold starts. The spark-plug tip may be wetted with unburnt or partially burnt fuel, or coated with soft, fuzzy carbon. As the mixture becomes leaner, temperatures rise and the spark plug enters the self-cleaning zone. Should the mixture be leaned out further, the spark plug will overheat and become an auto-ignition source.* www.briskusa.com

The manufacturer's recommended heat range is nearly always the best choice. However, engines that have been modified or those that see service in extremely cold weather may benefit from a one-step colder spark plug. The same holds when using liquefied petroleum gas (LPG) fuel. Lead replacement fuel (LRP), available from most service stations in Australia,

FIGURE 3-4. *The darker the spark-plug tip, the richer the mixture.*
U.S. Bosch

FIGURE 3-5. *As temperatures rise, the insulator tip whitens and the sharp edges of the electrodes round off. An air leak downstream of the carburetor or a clogged or improperly adjusted main jet are likely causes.* U.S. Bosch

burns cooler than conventional gasoline and so requires a slightly hotter spark plug to burn off the residue.

It's obvious that the replacement spark plug must be physically compatible, that is, have the same thread diameter and reach (thread length) as the original. A single- or double-digit number in the plug designation indicates the heat range. As shown in Table 3-1, Bosch and American brands assign higher numbers to hotter plugs. A Champion LM19 is hotter than an LM17.

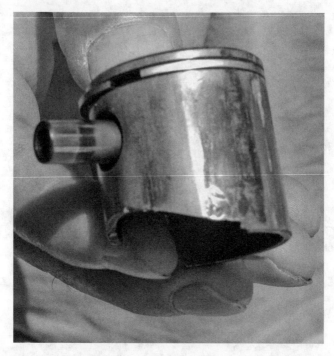

FIGURE 3-6. *Lean mixtures are death on engines, as evidenced by this battered and scarred piston.* Photo by Tony Shelby

TABLE 3-1. Heat range identification

Brand	Location of heat range numeral in model designation	Examples	Higher heat range number = hotter	Higher heat range number = colder
Autolite	Last digit	2974 4 = heat range	Yes	
Bosch	Middle digit	FR9DC 9 = heat range	Yes	
Champion	Middle one or two digits	J17LM 17 = heat range	Yes	
Denso	One or two digits after initial letter	W9LMR-US 9 = heat range		Yes
NGK	Middle one or two digits	BPR5EY 5 = heat range		Yes

Japanese manufacturers do the opposite: the higher the code number, the colder the plug. Nor do the heat range numbers correlate between brands. For example, a Denso IK20 has roughly the same heat range as a NGK BPR6IS or a Bosch FGR7KQE.

Repairing spark-plug threads

Stripped spark-plug threads in aluminum or cast-iron cylinder heads can be repaired with inserts, although heat dissipation may suffer a bit. Amazon's Save-A-Thread kit and the Fix-A-Thred cost about $30, but come only in the M14 1.25-mm size. Heli-Coil repair kits are available in M10, M12, M14, M18, and 7/8-in. sizes, but cost almost $200 each. Probably the most economical alternative for DIYers is to farm out the repair to an auto-parts store that does automotive machining.

Flywheel

It is necessary to remove the flywheel to access the crankshaft key and the under-flywheel magneto found on some vintage engines. Disconnect and ground the spark plug; remove the sheet-metal shroud and, on electric-start models, the starter motor. The flywheel brake found on rotary mowers can usually be left undisturbed.

The flywheel nut on modern engines has a standard—"righty tighty, lefty loosey"—thread. Some very early Briggs & Stratton horizontal-crank engines have a left-hand thread (Fig. 3-7). Most shops loosen the flywheel nut with an impact wrench. But an impact wrench cannot be trusted to tighten the nut. You can anchor the flywheel with a strap wrench (Fig. 3-8) or block the piston with nylon rope fed into the cylinder through the spark-plug boss (Fig. 3-9). A short length of two-by-four locks the blade on rotary mowers.

Once the flywheel nut comes off, remove the starter cup, noting how a tang on the underside of the cup indexes with the flywheel (Fig. 3-10). If a Belleville spring washer is installed between the cup and flywheel, the concave side of the washer goes against the flywheel. The crankshaft taper makes an interference fit with the flywheel that must be overcome with a gear puller or, more crudely, by shock. On some engines two or three tapped holes in the flywheel hub provide purchase for a gear puller (Fig. 3-11). European two-stroke flywheels have an internally threaded hub that requires a special puller to remove. These tools are hard to come by, although it's sometimes possible to find a bicycle crank-arm puller that works.

Caution: Do not attempt to remove a flywheel with a gear puller of the kind that hooks over the flywheel rim. While this procedure is permitted on engines with massive flywheels, the typical flywheel merely warps and grips the crankshaft tighter.

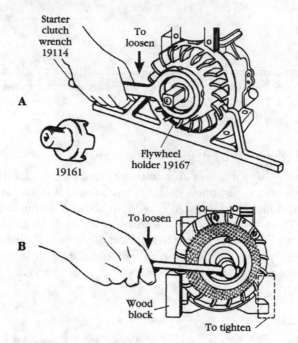

FIGURE 3-7. *Briggs engines with the classic "square-shank" rewind starter clutch (A) are still encountered. The special wrench goes for $20 or so on eBay, although in a pinch you can use a block of hardwood as a driver against the threaded "ears" on the clutch and a hammer. A socket wrench works for all other engines (B).*

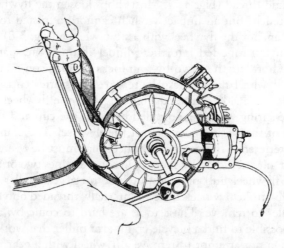

FIGURE 3-8. *A strap wrench will prevent the crankshaft from turning as the flywheel nut is loosened or tightened. But the flywheel brake on rotary mowers must be detached to provide clearance for the strap.*

FIGURE 3-9. *Rope can be used to block crankshaft motion.* Photo by Tony Shelby

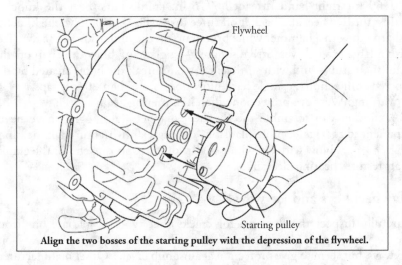

Flywheel

Starting pulley

Align the two bosses of the starting pulley with the depression of the flywheel.

FIGURE 3-10. *Starting pulley tangs index with the flywheel.*

FIGURE 3-11. *If the flywheel is drilled for puller purchase, an automotive harmonic balancer puller can substitute for the factory tool. Hub holes in Briggs & Stratton flywheels are unfinished and must be threaded with a 5/16-in. × 18 tap.* Photo by Tony Shelby

Other flywheels, including those on handheld two-stroke engines, must be shocked off. Most mechanics use a heavy brass bar placed squarely against the end of the crankshaft. Tecumseh used threaded knockers that can sometimes be found on EBay: PN 670105 for right-hand, 1/2-in. by 20 threads; 670118 for the vintage left-hand version of the same threads; and PN 670169 for right-hand threaded 7/16-in. crankshafts. Seat the knocker against the flywheel and back it off three turns (Fig. 3-12). Give the knocker a sharp rap with a hammer while prying up on the underside of the flywheel with a large screwdriver. Strike squarely—a glancing blow can snap off the crankshaft stub—and with sufficient force to break the bond created by the tapered connection. A single hard blow does more than 20 soft raps.

Warning: Wear eye protection when hammering against steel.

A knocker can scramble the flywheel magnets (although I have never experienced that) and, more plausibly, dislocate anti-friction main bearings (Fig. 3-13). A blow with a rawhide mallet on the opposite end of the crankshaft restores bearing clearances.

Flywheel keys and keyways

Carefully inspect the flywheel for cracks that typically radiate out from the keyway (Fig. 3-14). Briggs uses soft aluminum keys that wallow—out keyways, but seem to give protection against hub cracks. Other manufacturers use steel keys.

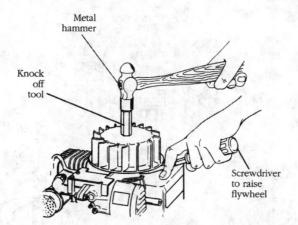

FIGURE 3-12. *Millions of flywheels have been successfully knocked off, but the procedure entails risk. Back off threaded knockers three turns from seated and strike squarely, as if driving a nail.*

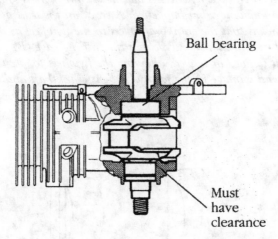

FIGURE 3-13. *Shocking a flywheel loose can displace anti-friction main bearings in their housings. Correct by striking the opposite end of the crankshaft with a mallet.*

Warning: Always replace cracked flywheels. Hub cracks propagate until a critical length is reached and the flywheel grenades.

Remove the flywheel key from the crankshaft stub with side-cutting diagonal pliers ("Dykes"). A few thousandths of an inch of key wallow or distortion, multiplied by flywheel diameter, has a major effect on ignition timing (Fig. 3-15). A sheared key denies ignition.

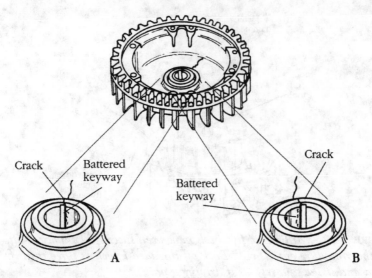

FIGURE 3-14. *A cracked flywheel is bad news, but there is some consolation in knowing what happened. A crack on the leading edge of the keyway (A) means that the crankshaft overran the flywheel as the result of a loose flywheel nut and taper. A crack on the trailing edge (B) suggests that the crankshaft stopped or slowed, allowing the flywheel to overtake it. Expect to find collateral damage in the form of a bent crankshaft and/or rotary-mower blade.*

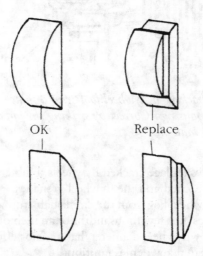

FIGURE 3-15. *Replace distorted or worn keys.*

Engine makers recommend that the crankshaft and/or flywheel be replaced when keyways loosen. New parts would be certain to restore original timing, but the correct timing depends upon keyway alignment. The flywheel key is merely a convenience. With care a flywheel can be mounted without a key and the engine will function normally.

Flywheel magnets can weaken in service or, as more often happens, when flywheels are stacked on top of one another in storage. Healthy flywheel magnets will attract a loosely held screwdriver through an air gap of 5/8 in. Smaller magnets used to time the spark on solid-state ignitions need not be that strong.

Flywheel installation

Clean corrosion off the flywheel and crankshaft tapers. Test-fit the flywheel on the crankshaft. It should drop down hard enough to stick. If a friction lock cannot be achieved, one or the other part is out of spec and should be replaced or, if you have the patience, reformed to fit with the help of Prussian blue and valve-grinding compound. It's good also practice to dummy up the assembly loosely and spin the flywheel by hand to verify that the ignition system delivers a spark.

If all is in order, replace the key and assemble the flywheel. As mentioned earlier, Belleville spring washers go on with the concave side against the flywheel. Tighten the flywheel nut to the specified torque limit. Table 3-2 lists the specification for some popular American utility engines.

Breaker-point magnetos

The unit shown in Figure 3-16 has all its parts grouped under the flywheel, which has magnets cast into the inner side of its rim. Other magnetos mount externally to the flywheel and are energized by outward-facing magnets. Early magnetos employed mechanical switches known as breaker, or contact, points. Modern magnetos use solid-state switches, but otherwise operate on the same principles.

The ignition coil is made up of two electrically independent windings wrapped over a laminated iron armature. The primary winding consists of some 200 turns of relatively heavy wire wound directly over the armature (Fig. 3-17). One end of the primary winding grounds to the armature; the other end grounds when the contact points or solid-state switch closes. The secondary winding consists of approximately 20,000 turns of hair-fine wire wound over the primary, but insulated from it. One end of the secondary shares the same ground as the primary, and the other end terminates in the spark-plug lead. When the spark plug fires, the circuit completes itself to ground.

TABLE 3-2. Flywheel nut torque for selected four-stroke engines

Make	Model	Flywheel nut torque (lb-ft)
Briggs & Stratton	6000 to 13000 alum. block, L-head	55
	17000 to 25000 alum. block, L-head	65
	85400, 115400, 117400, 118400, 138400,	45
	97700, 99700, 12000, 18540, 235400, 245400	60
	200000, 210000, 28000, 310,000	100
Kohler	K91	40–50
	K161, K181	85–90
	K241, K301, K321, K341	50–60
	Magnum 8-16 hp	85–90
	Command	50
Tecumseh	VS, TNT, ECV, H/HSK, HS/HSSK, LEV, OHH, OVRM	42–50
	VM125, 140, H50-60, V70, H70	45–60
	TVM/TVXL 170M 195, 220 and HM/ HMSK70-100, OVM/OVXL, OHV120-125 OHSK80-130, OHM120, OHV11-13, OHV11-13, OHV110-135, 206, 203, OHV135-145, OHV 15-17,5, 204	78–90

As the magnetic field of force passes the ignition coil, the primary windings cut the lines of magnetic force, inducing a current flow in the primary winding.

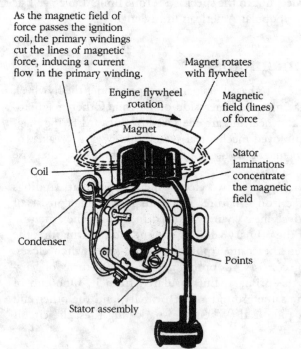

Magnet rotates with flywheel

Engine flywheel rotation

Magnetic field (lines) of force

Magnet

Stator laminations concentrate the magnetic field

Coil

Condenser

Points

Stator assembly

FIGURE 3-16. *An under-flywheel Phelan magneto. Elongated mounting slots permit the magneto to move a few degrees relative to the flywheel to vary the static timing. The drawing omits the breaker-point cam.*

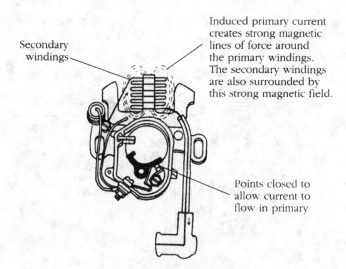

Secondary windings

Induced primary current creates strong magnetic lines of force around the primary windings. The secondary windings are also surrounded by this strong magnetic field.

Points closed to allow current to flow in primary

FIGURE 3-17. *With the points closed, the primary circuit is complete and coil windings are saturated with magnetic flux.*

Figures 3-17 and 3-18 illustrate magneto operation. As the flywheel turns, a magnet sweeps past the coil to produce voltage in the primary winding. When a conductor is exposed to a moving magnetic field, voltage is induced in the conductor. So long as the points remain closed, both ends of the primary winding are grounded to complete the circuit, and current flows.

Primary current creates a strong magnetic field in the secondary windings. Further movement of the crankshaft opens the points to deny ground to the primary circuit. Current flow ceases, and the magnetic field that permeated the secondary windings falls inward upon itself. The sudden collapse of the magnetic field induces 15,000 V or more in the secondary windings. This voltage finds release by arcing to ground across the spark-plug electrodes.

The "hot" side of the condenser ("capacitor" in modern terminology) connects to the primary winding, usually by way of the movable point arm. The other side of the condenser grounds to the engine through the metal case. When the points break open, the condenser charges with electrons that would otherwise seek ground by arcing across the point gap. Milliseconds later, the primary voltage diminishes enough to permit the condenser to discharge through the primary winding. The backflow of electrons from the condenser bucks primary voltage to speed the collapse of the magnetic field.

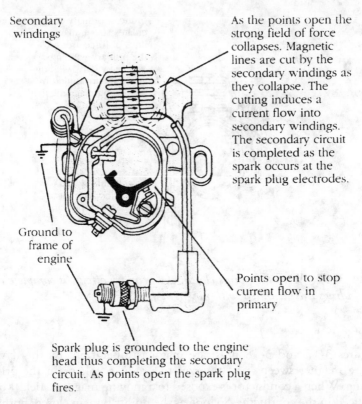

Secondary windings

As the points open the strong field of force collapses. Magnetic lines are cut by the secondary windings as they collapse. The cutting induces a current flow into secondary windings. The secondary circuit is completed as the spark occurs at the spark plug electrodes.

Ground to frame of engine

Points open to stop current flow in primary

Spark plug is grounded to the engine head thus completing the secondary circuit. As points open the spark plug fires.

FIGURE 3-18. *When the contact points cam open, the primary circuit opens and the surrounding magnetic field implodes to induce high voltage in the secondary windings.*

Servicing contact-point magnetos

First verify that the flywheel key is undamaged and check that the kill switch wiring is not shorted. Set the point gap—usually 0.020 in.—and install the flywheel without torqueing down the nut. With the spark plug removed, spin the flywheel. There should be a spark. If not, replace the point set and condenser. You can sometimes salvage a point set by filing it (never use an abrasive, which contaminates the tungsten contacts). Burnish the new contacts with a business card to remove fingerprints. You may have to narrow the point gap to get a spark on a tired magneto. If there is still no spark, replace the ignition coil or, when practical, replace the magneto with a more modern spark generator.

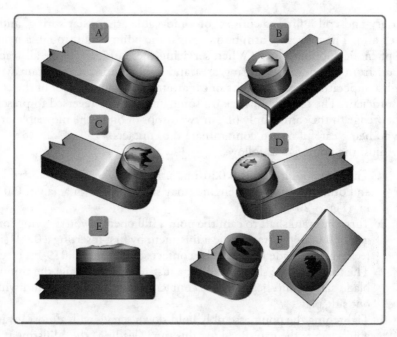

FIGURE 3-19. *Contact point maladies.*

Worn points tell a story. "Frosted" contacts crystallize from heat, which also blackens the supporting arm (Fig. 3-19A). The cause is arcing from an open condenser. Dirt-contaminated points (Fig. 3-19B, C, and D) have minor pits that over time coalesce into a central crater. The blackened contacts show evidence of overheating. Metal transfer between contacts (E) occurs because of a bad condenser or a loose wire between the point set and the coil. Adjacent surfaces remain bright with no evidence of overheating. Sooty points (E) result from oil contamination.

Oil fouling is the result of wear on the flywheel-side crankshaft seal or because of oil seepage around the plunger used on some Briggs and Kohler engines. This condition may be accompanied by a splatter of carburized oil under the contacts.

The contacts should meet over their full diameter (Fig. 3-18). Correct any misalignment by bending the *fixed* arm.

"Standard" breaker point sets

Point assemblies come in two varieties. The standard arrangement found on Wico, Phelon, Bosch, Lucas, and Bendix magnetos and on all battery-and-coil ignitions consists of a moveable arm, a flat spring, and a fixed arm. The moveable arm bears against the point cam through a nylon or phenolic rubbing block. When servicing these units, remove all traces of oil from the point mounting area and lightly lubricate the cam with high-temperature grease. A light smear around the full diameter of the cam is sufficient. The oil-wetted wick on some units can be reversed to present a new edge to the cam. Apply one or two drops of oil to the moveable-arm pivot, being careful not to contaminate the contacts.

Adjust the point gap as follows:

1. The basic point gap is 0.020 in., although West Bend two-strokes and other really ancient engines may start easier with a gap of 0.018 in. (Fig. 3-20).
2. Turn the crankshaft to cam the points full open. If you're trying for a gap of 0.020 in., bracket the adjustment by first inserting a 0.021-in. feeler-gauge blade between the contacts and then a 0.019-in. blade. The thicker blade should exhibit a just-perceptible drag; the thinner blade should slip between the contacts without touching and without side play.
3. Tightening the point assembly hold-down screws will, almost invariably, change the gap. Readjust this time "leading" the adjustment to compensate for hold-down distortion. Patience is required.
4. Burnish the contact faces with a business card to remove fingerprints, oxidation, and any oil that may have transferred from the feeler-gauge blades.

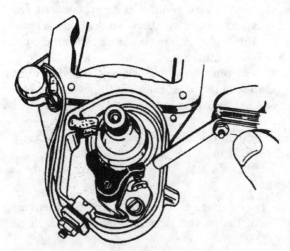

FIGURE 3-20. *Gapping a "standard" (grounded fixed arm) point set.*

5. Install a new condenser. Wipe up any oil from the condenser seating area—which should have zero electrical resistance—and keep the condenser lead clear of the flywheel. Verify that the moveable arm spring has not twisted into contact with block metal.

Briggs & Stratton point sets

The most often encountered Briggs magnetos have a grounded moveable arm and a "hot" fixed arm integral with the condenser.

To service:

1. Remove the flywheel, point cover, and point assembly (Fig. 3-21).
2. Note the lay of the parts:
   a. The braided ground strap loops over the post that secures the moveable breaker arm.
   b. The open end of the point spring feeds through the larger hole in the breaker arm and exits through the smaller hole.

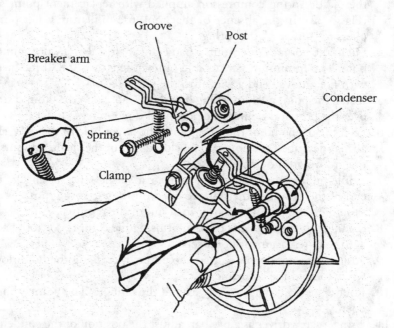

FIGURE 3-21. *Briggs & Stratton used this point configuration on light- and medium-frame engines. Note how the point spring mounts and the way the ground strap loops over the breaker-arm post.*

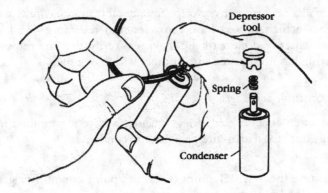

FIGURE 3-22. *Small-engine mechanics should have a Briggs spring compressor in their toolbox.*

3. Remove the coil and kill-switch wires from the condenser with the plastic spring compressor supplied with replacement point sets (Fig. 3-22). In the absence of that tool, use miniature water pump pliers to "unscrew" the spring.

4. Oil in the point cavity means a bad crankshaft seal or a worn plunger bore. The minimum permissible plunger length is 0.870 in. Install with the grooved end adjacent to the point set.

5. Install the point port, indexing it with the tab. Loop the ground strap over the post and tighten the post hold-down screw.

6. Feed the open end of the spring into the larger hole in the moveable arm as described previously. The closed end of the spring mounts in the groove on the smaller post.

7. While pulling the moveable arm against spring tension, engage the end of the arm into the slot provided on the mounting post. Wires should extend about a quarter inch out of the terminal.

8. Connect the wires to the condenser using the compressor tool packaged with replacement point sets.

9. Position the condenser so that the points met. Snug the hold-down screw to hold the adjustment.

10. Rotate the crankshaft to fully extend the plunger. The points should open.

11. Using a screwdriver as a pry bar, adjust the position of the condenser to provide a 0.020-in. point gap.

12. Torque the hold-down screw and recheck the point gap (Fig. 3-23). Burnish the contact faces with a business card.

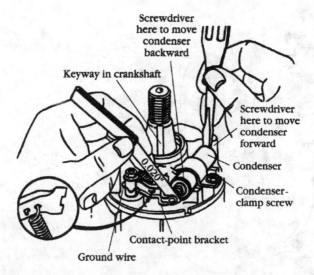

FIGURE 3-23. *To adjust Briggs & Stratton points, turn the crankshaft to bring the keyway into alignment with the point plunger, snug down the condenser clamp screw, and set the gap at 0.020 in. Tighten the screw and recheck the gap.*

Magnetron solid-state magnetos

Magnetron magnetos, standard on contemporary Briggs & Stratton engines, substitute a transistorized switching circuit for the troublesome point set. These far more reliable magnetos can be recognized by a single wire running from the coil to the kill switch (Fig. 3-24); point-set Briggs magnetos have a kill-switch wire and a second wire to the under-flywheel point set.

Operation

A trigger coil piggy-backed on the ignition coil generates voltage as the flywheel magnets come into proximity with it (Fig. 3-25). This voltage causes the Darlington transistor pair to complete the primary circuit to ground. Further movement of the flywheel induces a 3-ampere (A) current in the primary that saturates the secondary winding with magnetic flux.

As the flywheel continues to turn, magnetic polarity reverses to cause a similar reversal in trigger-coil voltage. The Darlington transistor switches off. Denied ground, primary current ceases to flow and the magnetic field surrounding the secondary windings collapses at near light speed. This collapse induces high voltage in the secondary circuit. The spark plug fires.

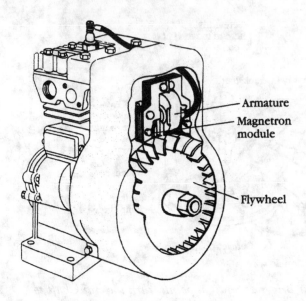

Figure 3-24.
Magnetrons mount outside of the flywheel and have a single wire going to the kill switch.

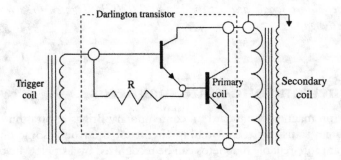

Figure 3-25. *The Magnetron trigger circuit is similar to that used in automotive systems.*

Ignition timing advances linearly with engine rpm. At low speeds the flywheel magnets must be within close proximity to the trigger coil to induce the 1.2 V needed to activate the transistors. Higher speeds lower the magnetic threshold and ignition occurs earlier.

Resistance readings between the spark-plug terminal and an engine ground should be between 3000 and 5000 ohms. Early production units had replaceable trigger modules, which was a good feature since the transistors are subject to heat damage. The factory stamps Magnetron coils with the date of manufacture, which will be a month or so earlier than the engine build date.

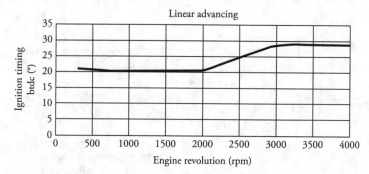

FIGURE 3-26. *Subaru solid-state ignitions deny advance at low engine speeds.*

More sophisticated solid-state magnetos, such as those used by Kohler and Subaru, have a low-speed circuit that retards ignition advance during cranking to reduce kickback (Fig. 3-26). In the distant past, motorists who crank-started their cars risked a broken arm if they forgot to retard the ignition.

Aftermarket upgrades

Several kits are available to upgrade point-and-condenser magnetos to solid-state operation:

- Briggs & Stratton Magnetron kits (PN 394970) are designed to work with Briggs two-legged coil armatures.
- Mega-Fire and Nova 2 kits are nearly universal in that they are compatible with two- and three-legged armatures and magnets on inner and outer flywheel rims. But because each pass of a magnet produces a spark, these units are incompatible with multiple-magnet alternator flywheels.
- Kohler PN 25 757 10-S can be used on many, but not all other engines. Depending upon where you buy the unit, costs range from $35 to $50.

The kits described above systems are powered by alternating current generated by the flywheel magnets. Brian Williams supplies kits, using solid-state automotive ignition components, for points-and-condenser systems that run off direct-current battery power.

CDI systems

Capacitive discharge ignitions (CDIs), originally developed for racing, are almost universally used on chainsaws, outboard motors, and other high-revving small engines. CDIs generate higher open-circuit voltage

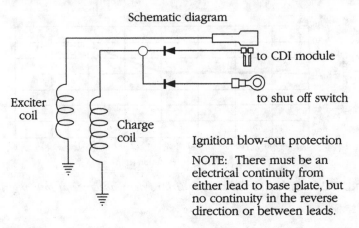

Schematic diagram

Exciter coil

Charge coil

to CDI module

to shut off switch

Ignition blow-out protection

NOTE: There must be an electrical continuity from either lead to base plate, but no continuity in the reverse direction or between leads.

FIGURE 3-27. *CDIs use diodes to convert alternating current to direct current for charging the capacitor.*

than magnetos, and the voltage builds more rapidly—in some examples 10,000 V/μs (volts per microsecond). However, most four-stroke utility engines continue to use solid-state magnetos, which are simpler, less expensive, and probably more reliable than CDIs. The long spark duration characteristic of magnetos also assists in firing fouled spark plugs.

CDI systems charge a capacitor from a flywheel coil or, on some larger engines, with a 12-V battery. A coil generates alternating current: output current reverses polarity as the flywheel magnet retreats. In order to charge the capacitor, most CDIs employ blocking diodes—the electronic equivalent of check valves—to block current flow in one direction (Fig. 3-27). Another approach is to arrange matters so that the capacitor reaches full charge before the current reversal.

Precautions to take with CDI systems

CDI systems are sensitive to reversed polarity and stray voltages:

- If the circuit includes a battery, the negative (black) terminal goes to ground. Reversing polarity is death on the CDI system unless the circuit includes blocking diodes.
- Do not run the battery systems with the battery disconnected. The battery acts as a voltage-limiting resistor.
- Do not leave a spark-plug lead disconnected and hanging loose when test cranking the engine. Open-circuit CDI voltage can puncture coil insulation.
- And finally, do not introduce stray voltages by welding on the driven equipment with the CDI installed.

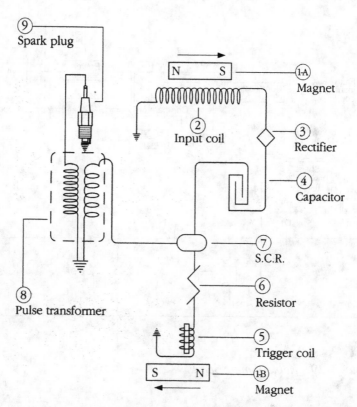

FIGURE 3-28. *Tecumseh CDI. This system advances ignition as a function of flywheel-magnet velocity. At low engine speeds, the magnetic field created by the flywheel magnets moves relatively slowly. Trigger coil excitation must wait until the flywheel is in close proximity to the coil. At higher engine speeds, the more rapidly moving field generates trigger coil voltage earlier, while the flywheel is at some distance from the coil.*

As with other types of ignitions, check the peripherals—the flywheel key, spark plug(s), external wiring, and interlocks—before assuming that the CDI has failed.

Tecumseh CDI

Figure 3-28 is a schematic of the basic circuit. The flywheel magnets (1A) induce 200 VAC in the input coil (2). The rectifier (3) converts the output to direct current for storage in the capacitor (4), and a silicone-controlled rectifier (SCR) remains nonconductive to block capacitor discharge. At approximately 180° later of crankshaft rotation, the flywheel magnets sweep past the

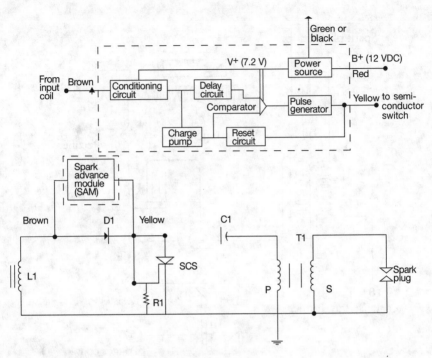

FIGURE 3-29. *Kohler Smart Spark.*

trigger coil (5) to generate a signal voltage across the resistor (6). This voltage causes the SCR to conduct. The capacitor (4) then discharges into the primary side of the pulse transformer (8). Current flow in the primary windings of the transformer—actually an ignition coil—generates a 25,000-V potential in the secondary windings that goes to ground through the spark-plug electrodes. Ignition advance is linear with engine rpm. Other CDIs retard the spark for easier starting (Fig. 3-29), and many include an rpm limiter.

Smart Spark

Kohler's Smart Spark CDI advances the spark in a nonlinear manner with engine speed. As mentioned earlier, all solid-state ignitions have some built-in advance capability, since an increase in flywheel speed induces voltage earlier in the trigger coil. The Kohler unit quantifies the advance to more accurately reflect engine requirements.

Figure 3-29 sketches the main features of the Smart Spark. Flywheel magnets induce AC voltage in the input coil (L1). Part of the coil output passes through the rectifier diode (D1) and charges the main capacitor (C1).

A smaller fraction of the coil output goes to the spark advance module (SAM) mounted externally on the engine shroud. This conditioning circuit shapes the power pulse, which then goes to what Kohler calls the charge pump. The "pump" charges the main capacitor in proportion to engine speed.

Both the charge pump and the delay circuit include capacitors. When the charge on the delay circuit capacitor exceeds the charge on the charge pump capacitor, the comparator fires the semi-conductor switch (SCS) to discharge the main capacitor. The resulting high voltage in the coil secondary (S) finds ground by arcing across the spark-plug electrodes.

Ignition timing depends upon the time required for charge pump and delay circuit capacitors to charge. At low engine speeds, the delay circuit capacitor is slow to charge and ignition is delayed. As speed increases, the charge builds faster and the spark occurs earlier. The trigger pulse from the SAM charges the capacitor in the reset circuit, clearing the decks for the next revolution of the crankshaft.

Troubleshooting the Smart Spark

As with other ignition systems, replace the spark plug and, if damaged, the flywheel key. Make a careful examination of the external wiring, looking for loose connections, bad grounds, and chaffed insulation. In some applications these external circuits draw enough power to make starting difficult. The SAM needs 7.2 V to function. Verify that the battery has a full charge, and make resistive checks of the ignition switch and associated wiring.

When a Smart Spark fails, the problem becomes one of deciding which module—ignition or SAM—to replace. Resistance tests of the ignition module, although less than definite, are helpful. With the module at room temperature, disconnect the brown wire and measure the resistance between the wide connector tab and the coil laminations. Resistance should be between 145 and 160 ohms. Remove the yellow lead and measure resistance from the narrow tab to the laminations, which should be 900–1000 ohms. The resistance from the spark-plug terminal and the laminations should be 3800–4400 ohms. Large discrepancies from these figures suggest that the ignition module has failed. Otherwise, replace the SAM.

Setting the E-gap

The armature air gap, or E-gap, is the distance the coil armature stands off from the flywheel rim. The narrower the gap, the stronger the magnetic field and the more voltage is induced in the coil. However, some clearance is necessary to accommodate main-bearing wear, flywheel growth at high rpm, and production variations. Skid marks on the flywheel rim mean that the E-gap is too narrow.

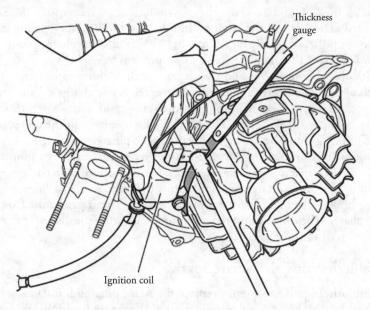

FIGURE 3-30. *Setting the E-gap on a Robin engine with a nonmetallic feeler gauge.*

The recommended E-gap falls within the range of 0.006 to 0.012 in. While nonmagnetic feeler gauges exist, many mechanics use a business card (Fig. 3-30).

Procedure:

1. Loosen the armature hold-down screws.
2. Rotate the flywheel to bring the magnets adjacent to the armature.
3. Insert the business card or nonmagnetic feeler between the flywheel rim and armature.
4. Tighten the hold-down bolts.
5. Rotate the flywheel to retrieve the gauge.
6. Turn the engine over several times to verify that the clearance between the flywheel rim and armature is never less than 0.006 in.

Battery and coil ignitions

Battery-and-coil ignitions found on older Kohler, Onan, and Wisconsin utility engines and on a variety of larger power plants have several advantages. Not the least of these advantages is that power is constant and easily traceable.

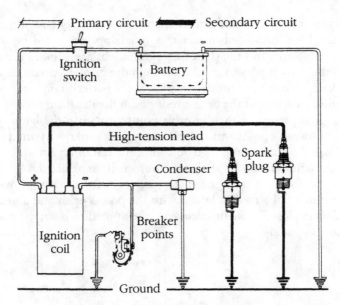

FIGURE 3-31. *Battery-and-coil ignition formerly used on twin-cylinder Kohlers.*

Operation

Figure 3-31 illustrates the wiring diagram of a four-cycle, twin-cylinder engine. This example lacks a distributor, which means that both cylinders fire simultaneously, with the piston in one cylinder on the compression stroke and the other piston on exhaust stroke. The "phantom" spark is also present on four-cycle engines that take their firing signals from the crankshaft.

The primary side of the circuit consists of a 12-V battery; ignition switch; points and condenser; a resistor, integral with the coil or in the form of a stand-alone ballast resistor; and primary coil windings. The secondary coil winding connects to the spark plugs through high-tension leads. Both the primary and secondary ground to the engine.

The system functions like a magneto, except that battery voltage, rather than magnetic induction, excites the primary winding.

Troubleshooting

Most problems with the system result from a weak battery. Point sets (including the condenser) are the next most likely suspect. Note that Chevrolet small-block point-and-condenser sets substitute for the Kohler parts. The point gap for all applications is 0.020 in.

With the switch "On" and points open, check primary circuit continuity with a test lamp or volt-ohm-meter. A test lamp connected between the point movable arm and the stationary (grounded) arm should glow dimly. Some of the voltage goes to ground through the primary circuit. If no voltage can be detected, trace the circuit back to the battery. Loose connections, the ignition switch, and the ballast resistor are the usual culprits.

If voltage is present at the moveable point arm when the lamp is placed across the moveable point arm and stationary arm, turn the flywheel until the points close. The lamp should go out, since the stationary arm is grounded. Should the lamp remain lit, the point contacts have oxidized.

When the results of these tests are negative—that is, when primary voltage is present on the moveable arm with the points open and absent when the points are closed—the problem can be assumed to be in the condenser or the coil secondary.

4

Fuel systems

Most small-engine malfunctions that a new spark plug does not cure originate in the fuel system. This system consists of the fuel tank, pulse and fuel lines, carburetor, and engine induction tract.

Tools and supplies

You will need a clean work space large enough to lay out parts in order and:

- Flat-blade and Philips screwdrivers.
- Long-nosed pliers.
- Turkey baster to empty fuel tanks.
- Mineral spirits or another solvent as described in the "Carburetor Cleaning" section.
- Steel wool for removing rust and gum from float bowls.
- Pipe cleaners.
- Paper towels or lint-free rags.
- Gloves. Nitrile gloves provide good protection against gasoline, oil, and grease, but they offer only what is described as "fair" protection against acetone, which is a primary ingredient in carb cleaners.
- Safety glasses, especially if aerosol cleaners or compressed air is used.

Serious mechanics need:

- Tecumseh PN 670377 inlet valve seat removal and pilot tool for recent float-type carburetors (retails for $8 to $15).
- Walbro 50-13-1 and ZT-1 Zama metering-lever gauges for diaphragm carburetors (about $8).

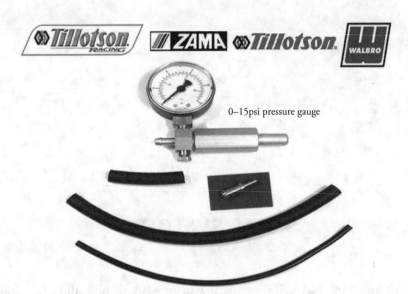

0–15psi pressure gauge

FIGURE 4-1. *Mowers4u supplies this pressure tester, identical to those used by OEMs. Walbro stocks a rebuild kit for the pump.*

- Mixture adjustment tools for diaphragm carburetors. Most common sizes are Zama 21-Teeth Splined, Pac Man, Single D, Double D, Hex Socket (around $20). Some Walbro diaphragm carburetors require a 22-Tooth Splined tool.
- An OEM (Walbro, Zama, Tillotson) pressure gauge and fittings available from Mowers4u for about $60 (Fig. 4-1) or purchase a $35 Mityvac MV8000 vacuum/pressure pump.

Fuel tank

Check for leaks and, on steel tanks, for rust that has become a much more serious problem with the advent of ethanol-blended gasoline. Fuel tank vents for handheld engines close when the tank is inverted and open to relieve excessive vapor pressure. If the engine shuts down after a few minutes of operation, loosen the fuel-tank cap. Should that solve the problem, the vent is clogged and needs cleaning or replacement.

Fuel filters

Filters for four-stroke engines mount between the fuel tank and carburetor. Install them with the arrow indicating flow direction oriented correctly.

Handheld engines have in-tank filters extracted through the filler hole with a bent wire. If the filter is deeply stained, replace it.

Fuel and pulse lines

Pulse lines that transfer crankcase vacuum to the fuel pump tend to clog, as does the crankcase pulse port. Visually inspect rubber fuel hoses for cracks and leaks. Plastic lines, used to transfer fuel and crankcase pressure pulses for handheld tools, become brittle in service and can develop soft spots that collapse to deny fuel delivery. To pressure test the plumbing:

- Disconnect the fuel line at the carburetor inlet fitting.
- Clamp off the line with Vise-Grips.
- Disconnect the filter and extract the line from the tank.
- Connect a Mowers4u or Mityvac pressure tester to the open end of the line.
- Pressurize to 10 psi maximum. If pressure holds for one minute, the lines are in good shape.

Most handheld tools use 0.110-in. inner diameter (ID) by 0.190-in. outer diameter (OD) or 1.040-in. ID by 0.230-in. OD plastic line. To install a new fuel line, lubricatei it with motor oil and press it through the tank gasket. Remove the gas cap and use long-nosed pliers to pull several inches of the line out of the tank. Install the filter and filter clamp. Upon assembly, check for fuel leaks and recheck after the engine has run for 20 minutes.

Air filters

Modern four-cycle engines have pleated paper filters that, when operating properly, capture dust particles as small as 10 microns in diameter. (One micron equals one-millionth of a millimeter or four one-hundred thousandths of an inch.) Figure 4-2 illustrates the paper filter used on Wisconsin engines.

Other than periodic replacement, paper capture filters require no service. Opening the canister for inspection introduces dust into the carburetor. Attempting to clean the filter with compressed air opens voids in the paper that can be seen as streaks of orange when the filter is held up to the sun. Liquids, including gasoline from flooded carburetors, swell the fibers into impermeability.

Current practice is to encase the paper element with a polyurethane foam "sock" that catches larger dust particles. These prefilters need periodic

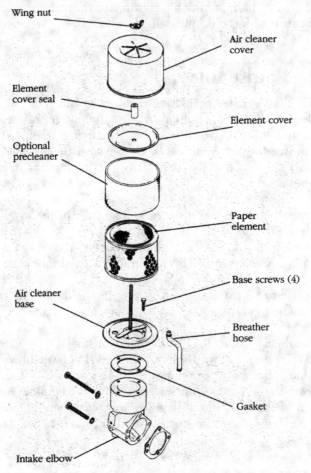

Wing nut

Air cleaner cover

Element cover seal

Element cover

Optional precleaner

Paper element

Base screws (4)

Air cleaner base

Breather hose

Gasket

Intake elbow

FIGURE 4-2. *A Wisconsin air cleaner combines a pleated paper element with a polyurethane foam prefilter. When installing a new prefilter, turn it inside out to make a good seal with the paper element.*

cleaning with hot water and detergent—not gasoline. Rinse with water, allow to dry, and dampen the foam with a light coating of motor oil. Excessive oil will damage the paper filter element.

Diaphragm fuel pumps

Engines with float-type carburetors are sometimes fitted with remotely mounted diaphragm fuel pumps (Fig. 4-3). One side of the diaphragm sees

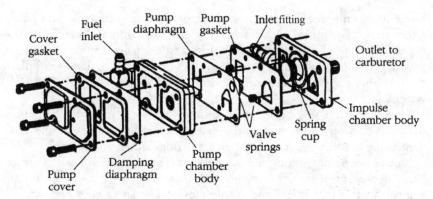

FIGURE 4-3. *Briggs & Stratton fuel pump includes a second diaphragm to smooth output pressure pulses.*

crankcase pressure fluctuations, and the other side opens to the atmosphere. To test the pump, loosen the output line and crank the engine with the ignition Off. Catch the discharged fuel in a container and wipe up any spills.

Warning: Fuel leaks pose a potentially lethal fire hazard.

Repairs consist of replacing the diaphragm. Scribe mark the housings before disassembly and make note of the position of the cover gasket relative to the diaphragm.

Carburetors

The better four-stroke engines are equipped with float-type carburetors, two-stroke engines almost always have diaphragm carburetors, and inexpensive Briggs four-strokes often have suction-lift carburetors perched on top of the fuel tank.

Cleaning

In the past we immersed carburetors overnight in Berryman Chem-Dip or the even more potent Bendix Metal Clean. Brass and aluminum parts came out sparkling. Modern carburetors with their nonremovable soft parts cannot tolerate these cleaners. Nor can our bodies. For example, Chem-Dip contains ethylene chloride, m-cresol, mixed xylenes, p-cresol, ethyl benzene, and hexavalent chromium, all of which are toxic. Long-term exposure can result in liver, kidney, and respiratory system damage. Hexavalent chromium is a known carcinogen.

Walbro suggests that mechanics use mineral spirits as an immersion cleaner, but petroleum-based solvents have little effect on varnish and gum.

Some health-minded DIYers dip carburetors in concentrated lemon juice, either straight or mixed with water. Original Clorox brand Pine-Sol works better than citrus juice or mineral spirits and does not appear to damage plastic parts. Pine-Sol works better and does not appear to damage to plastic or rubber parts. Soak the carburetor for a half-hour or so and rinse with water.

Marvel Mystery Oil cleans up aluminum oxide and molasses, although it can take a week or more to work, males rusty parts look almost like new. Tide is the best detergent for light grease, work clothes, and hands.

Out of necessity working mechanics spot clean carburetors with aerosol brake degreasers or carb/throttle body cleaners (Fig. 4-4). If you opt for an aerosol, wear a face shield and nitrile gloves. The spray goes everywhere. Work outside, if at all possible. Open windows do not provide proper ventilation, and fans merely circulate the vapor.

Many, probably most, mechanics clear clogs that the aerosol does not remove with a fine wire of the sort used for wire brushes and shipping tags. Carburetor manufacturers warn that this practice upsets calibration, but the effect does not appear to be catastrophic.

FIGURE 4-4. *Aerosol brake and carburetor cleaners perform as advertised, but they must be used with caution.* Photo by Tony Shelby

Unless the carburetor is extremely dirty, leave Welch (expansion) plugs and cups in place. Replacement plugs and installation instructions—especially critical for Zama and Tillotson units—come with factory overhaul kits (Fig. 4-5).

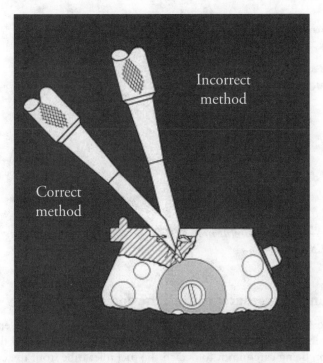

FIGURE 4-5. *Welch plugs are used to cover idle circuits and air bleeds after machining. Small cup plugs like those found on some Walbro diaphragm carburetors can be lifted with a No. 8 bottoming tap or a 5-mm set screw. Larger plugs are pierced with a small chisel and pried out. Hold the chisel at a shallow angle so as not to damage the fuel passages under the plug. Another and safer approach is to carefully drill the plug with a ¼-in. bit. Stop when the bit just breaks through, since there may be very little clearance between the plug and the backing material. Clean the fuel passages, and install a new Welch plug with the aid of a blunt tool. Flatten the plug even with the surrounding surfaces. These plugs install dry; ethanol gasoline dissolves fingernail polish and other sealants.* Used by permission of Walbro, LLC

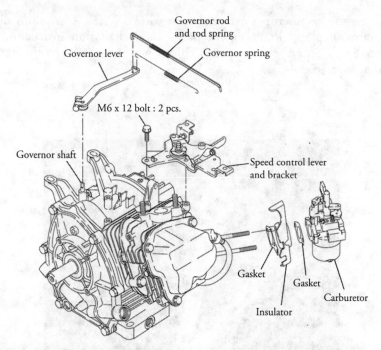

FIGURE 4-6. *Make careful note of linkages and linkage mounting points. Subaru EX shown.*

Detaching the carburetor from the engine

Close the fuel shutoff valve at the tank, or if a valve is not present, crimp the flexible fuel line closed with Vise-Grips. Disconnect the fuel line at the carburetor, being careful not to detach the fuel inlet tube from the carburetor body. Pressed-in inlet tubes sometimes pull loose as the hose is removed. A drop or two of blue Loctite 242 will secure the fitting in the future.

Remove the air-cleaner assembly, and, using long-nosed pliers, disengage the springs on the throttle/governor linkage. Spring tension and linkage attachment points are critical. Inspect the air-cleaner–to–carb-body gasket for damage. Complex governor linkages like the one shown in Figure 4-6 should be sketched or photographed before disassembly.

Disconnect the Bowen cable to the choke, if present. Back out the bolts that secure the carburetor flange to the engine block or inlet pipe or, on chainsaws, the clamps that secure the carburetor to the inlet hose. Once the carburetor is free, twist it to the right to detach the governor link from the throttle lever.

Inspect the carburetor flange gasket and heat shield/reed-cage gaskets for damage.

Carburetor basics

Carburetors work on the principle that nature abhors a vacuum. Air rushes into the carburetor bore, or throat, in response to the partial vacuum created by the motion of the piston. The carburetor bore resembles an hourglass, wide on the ends and narrow in the center (Fig. 4-7). The necked-down section is known as the venturi. As air passes through the venturi, its velocity increases and pressure decreases. The restriction, being curved, increases the distance the air must travel. Since as much air enters as leaves the venturi, the air passing through it must accelerate. And since nothing in nature is free, acceleration is purchased by a loss of pressure.

An aircraft wing demonstrates the same principle: air moving over the upper curved surface accelerates and loses pressure relative to the air passing over the flattened underside of the wing. The difference in pressure results in lift.

Fuel, stored under atmospheric pressure in the carburetor reservoir, moves into the partial vacuum created by the venturi, where it discharges in a mist of fine droplets. The more air passing through the venturi, the greater the vacuum and the more fuel discharged. Complications arise because the engine's appetite for fuel and air varies with operating conditions. The carburetor should supply seven parts of air by weight to one part fuel during cold starts. Acceleration requires an air/fuel ratio of 9:1; idling 11:1; and full throttle/full load operation 13.5:1 or richer.

Carburetors differ in the way fuel is introduced to the instrument. Float-type carburetors work like a toilet tank; diaphragm carburetors meter fuel entry with a flexible membrane; and suction-lift carburetors draw through a tube, as if drinking from a straw.

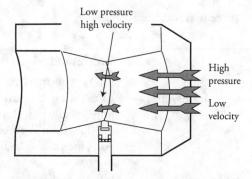

FIGURE 4-7. *A venturi, or any obstruction to airflow, such as a nearly closed throttle plate, results in reduced pressure and increased velocity.*

Float carburetor operation

Carburetors that maintain their internal fuel level with a float are the most common. The regulating mechanism consists of a float bowl, usually secured by a central nut, a plastic or metal float, and an inlet valve known as the needle and seat. As fuel is consumed, the float drops allowing the needle to fall away from its seat. Fuel enters the float chamber and the float rises to close the valve, an action that occurs several hundred times a minute at full throttle. Placing the fuel pickup tube in the center of the ring-shaped float enables the engine to run at shallow angles off the horizontal (Fig. 4-8).

High-speed circuit

At wide throttle angles the venturi creates enough vacuum to draw fuel into the pickup tube and out through the main nozzle (Fig. 4-9A). The main jet—a calibrated orifice in series with the pickup tube—regulates the rate of fuel discharge. An air bleed upstream of the main jet infiltrates the fuel with bubbles that assist in atomization and eliminate the potential for siphoning when the engine shuts down.

Low-speed circuit

A closed or nearly closed throttle plate acts like a crude venturi to generate low pressure downstream of its trailing edge. Idle-circuit ports, charged with fuel at atmospheric pressure, discharge into the depression. The port nearest the blade—the primary idle port—flows when the throttle plate rests against its stop (Fig. 4-9B). Secondary idle ports become active as the plate swings open (Fig. 4-9C).

Thus, we have three fuel sources:

Engine speed	Sources of fuel
Low idle	Primary idle port
Off-idle to one-quarter or one-third throttle	As the throttle opens, flow from the primary idle port diminishes and the flow from the secondary idle ports increases. The main jet begins to flow.
Between one-quarter or one-third throttle and wide-open throttle	Idle ports shut down and the main jet supplies all fuel.

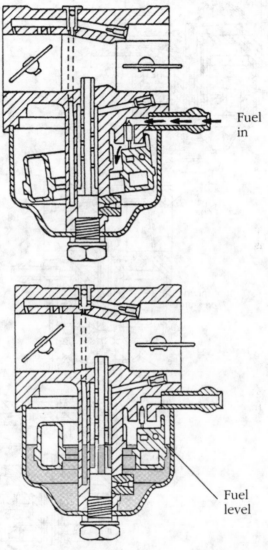

Fuel
in

Fuel
level

FIGURE 4-8. *The float mechanism maintains a preset level of fuel in the float chamber so that the air/fuel mixture is independent of the level of fuel in the tank.*

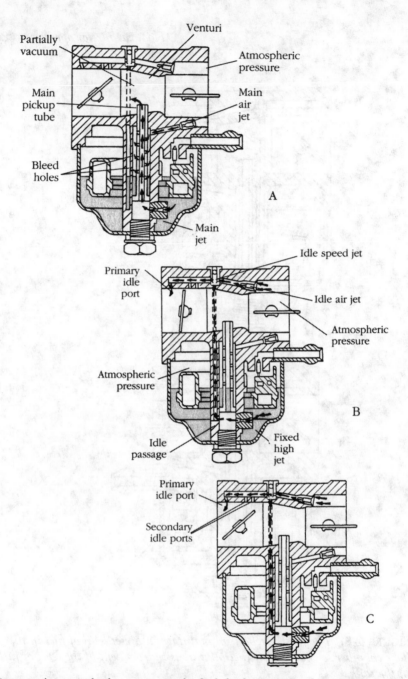

FIGURE 4-9. *At high engine speeds, fuel discharges into the venturi through the main nozzle (A). An air bleed emulsifies the fuel into droplets prior to discharge. During idle, the low-speed circuit discharges fuel from the primary idle port into the low-pressure zone downstream of the nearly closed throttle plate (B). As the throttle opens wider, secondary, or off-idle, ports supplement fuel delivery to smooth the transition to full throttle (C).* Used by permission of Walbro, LLC

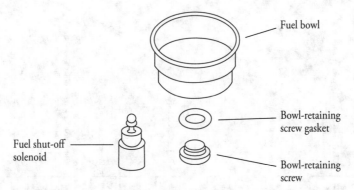

FIGURE 4-10. *A fuel shut-off solenoid, also called an anti-afterfire or anti-backfire valve, may be present under the float bowl.*

Fuel supplied to the idle ports passes through a metered orifice known as the idle jet. As far as consumers are concerned, neither the idle jet nor the high-speed jet is adjustable on modern carburetors. The only adjustment permitted is the idle-rpm screw on the throttle lever.

Anti-afterfire solenoid

When the ignition is switched off, the engine coasts down for a few revolutions before coming to a halt. If the throttle is open, fuel can pass through the combustion chamber and into the hot muffler, where it ignites with a bang. To prevent this, some float-type carburetors have an ignition-switch–controlled solenoid valve that cuts off fuel to the main jet. Figure 4-10 illustrates the valve used on Kohler engines.

Cold-start enrichment

Because gasoline requires heat to fully vaporize, more fuel must be ingested during cold starts. Until recently most small engines had a manually operated choke valve upstream of the venturi. With the choke closed, the engine pulls on what is essentially a blind pipe. The high level of vacuum draws fuels from all jets. More recent designs either make the choke action automatic or employ a manually operated primer pump.

Float carburetor service

Figure 4-11 pretty well describes all that can go wrong.

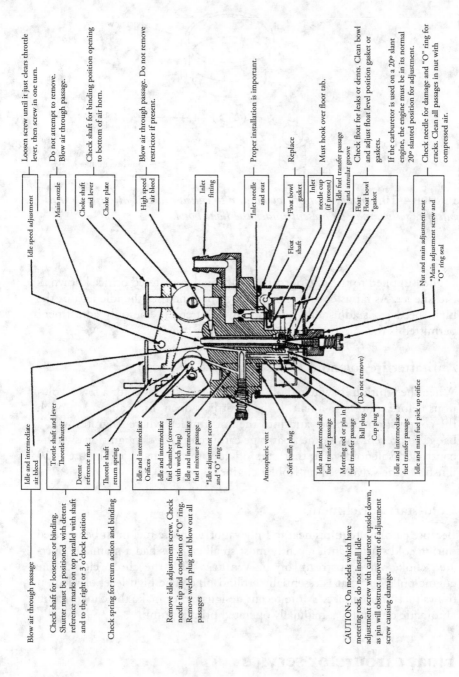

Idle and intermediate air bleed — Blow air through passage

Throttle shaft and lever Throttle shutter — Check shaft for looseness or binding. Shutter must be positioned with detent reference marks on top parallel with shaft and to the right or 3 o'clock position

Detent reference mark

Throttle shaft return spring — Check spring for return action and binding

Idle and intermediate Orifices

Idle and intermediate fuel chamber (covered with welch plug) Idle and intermediate fuel mixture passage

*Idle adjustment screw and "O" ring — Remove idle adjustment screw. Check needle tip and condition of "O" ring. Remove welch plug and blow out all passages

Atmospheric vent

Soft baffle plug

Idle and intermediate fuel transfer passage

Metering rod or pin in fuel transfer passage Ball plug (Do not remove) Cup plug — CAUTION: On models which have metering rods, do not install idle adjustment screw with carburetor upside down, as pin will obstruct movement of adjustment screw causing damage.

Idle and intermediate fuel transfer passage

Idle and main fuel pick up orifice

Idle speed adjustment — Loosen screw until it just clears throttle lever, then screw in one turn.

Main nozzle — Do not attempt to remove. Blow air through passage.

Choke shaft and lever — Check shaft for binding position opening to bottom of air horn.

Choke plate

High speed air bleed — Blow air through passage. Do not remove restrictor if present.

Inlet fitting

*Inlet needle and seat — Proper installation is important.

Float shaft

*Float bowl gasket — Replace

Inlet needle cup (if present)

Idle fuel transfer passage and annular groove — Must hook over floor tab.

Float Float bowl gasket — Check float for leaks or dents. Clean bowl and adjust float level position gasket or gaskets.

If the carburetor is used on a 20° slant engine, the engine must be in its normal 20° slanted position for adjustment.

Nut and main adjustment seat *Main adjustment screw and "O" ring seal — Check needle for damage and "O" ring for cracks. Clean all passages in nut with compressed air.

FIGURE 4-11. *Intended for Tecumseh carburetors, this drawing has application for all float-type carburetors.*

Fuel entry

The float bowl is secured by an anti-afterfire solenoid valve (recognized by the electric wire) or by a central nut that may double as the main jet. Some imported carburetors secure the bowl with multiple Phillips screws. The upper edge of the bowl seals against the carburetor body with a gasket or O-ring.

Anti-afterfire solenoid

A solenoid valve that sticks shut prevents the engine from starting or, if the sticking is intermittent, results in difficult-to-diagnose engine shutdowns. A valve that sticks open can send fuel dribbling out of the air cleaner or hydrolock vertical-shaft, four-stroke engines during storage. The latter condition occurs when gasoline collects in the cylinder and leaks past the piston rings to fill the crankcase. The crankshaft locks solid. In theory, the inlet needle and seat should prevent flooding, but needles and seats cannot be trusted to remain leak-proof during storage.

Warning: Gasoline-contaminated oil has the potential for explosion. If you detect the odor of gasoline on the dipstick, change the oil before starting the engine.

The solenoid should make audible clicks as the ignition switch is turned On and Off. The definitive test is to remove the device from the carburetor and, working in a well-ventilated place, free of gasoline vapor, connect the solenoid to a 12 V battery. It should snap open and closed without hesitation. If the action is desultory, try cleaning the plunger.

Solenoid valves cost between $40 and $80. You can neutralize a malfunctioning valve by grinding off the plunger with a Dremel tool and installing an inline fuel cutoff valve. Closing the throttle or putting a garden tractor/riding mower into gear before shutdown eliminates afterfire.

Warning: Inlet needle and seat valves in float-type carburetors must be backed by a manual or solenoid-operated shutoff valve. Otherwise the carburetor may leak fuel during periods of extended storage.

Float bowl

The amount of rust and sediment in the bowl is an indication of the condition of the carburetor. Tecumseh Series 7 and several Walbro carburetors have a cross-drilled float-bowl nut that functions as the main jet. Because of its vulnerable position, the jet quickly clogs.

Inlet needle and seat

Should the needle stick open, the carburetor quickly floods. Fuel runs out of the air cleaner and, when present, the external float-chamber vent. If the needle sticks shut, no fuel passes into the float bowl. It's worth repeating that needle and seat valves cannot be trusted to remain leak-proof during storage. The fuel supply should be contained with an manual shutoff valve or with an anti-afterfire solenoid valve. A leaking carburetor presents a potent fire hazard.

Older carburetors, units such as Walbro LMT, have replaceable brass needle seats and elastomer-tipped needles (Fig. 4-12). Take a careful look at the attachment hardware before removing the needle. Some of it is a bit tricky (Fig. 4-13). Leaks develop because of wear on the needle tip and as the result of grit impounded into the soft brass seat.

Many newer carburetors use an elastomer disc as a needle seat. Purchase several replacement discs since installation requires practice. Too little force results in leaks around the seat OD; too much force and the orifice distorts to prevent the needle from sealing. Figure 4-14 illustrates the service procedure.

Other later model carburetors use pressed-in brass inlet seats extracted with a drywall screw and Vise-Grips. Note how proud the original seat stands above the surrounding metal. Replacements come in two orifice sizes: the larger is for engines with fuel pumps. Using a small hammer and flat punch, install the new seat to the same depth as the original.

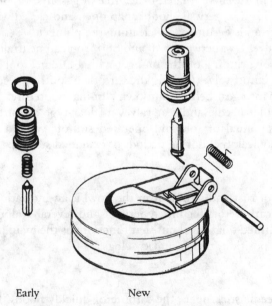

Early   New

FIGURE 4-12. *Needle and metallic seat assemblies.*

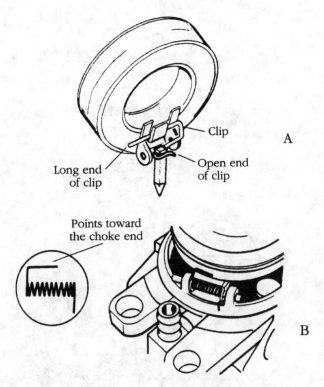

FIGURE 4-13. *Drawing (A) shows a version of a needle clip that should have its long end hooked to the float. Installed correctly, the dampening spring in drawing (B) exerts a slight lift on the float.*

Float-bowl vents

Modern carburetors are vented internally through a passage in the carburetor body that runs between the roof of the float chamber and the air cleaner. Externally vents open directly to the atmosphere, where they can more easily become clogged. Should this happen, the carburetor floods.

Float adjustments

The higher the float rises before seating the inlet needle, the richer the mixture. Metallic floats like those shown in Figures 4-15 and 4-16 have tangs that are bent to adjust float height and, on some examples, float drop. Height specifications vary; however, an engine will run if the float rests parallel to the float-chamber roof with needle and seat installed and the assembly inverted. Excessive drop can prevent the float from rising and cause massive flooding.

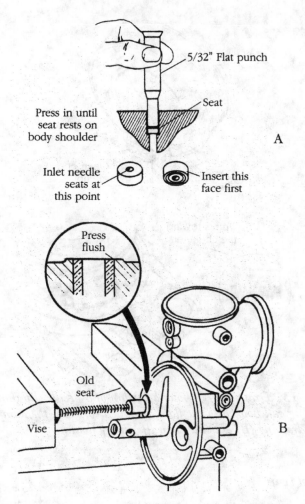

5/32" Flat punch

Seat

Press in until
seat rests on
body shoulder

A

Inlet needle
seats at
this point

Insert this
face first

Press
flush

Old
seat

Vise

B

FIGURE 4-14. *A hooked wire or dental pick can be used to extract the seat. Install with the grooved side down, toward the carburetor body. Tecumseh suggests that a punch be used for installation (A). Briggs seats are pressed in flush using the original as a cushion (B).*

Primer pump

Most modern float-type carburetors employ a primer pump instead of a choke plate (Fig. 4-17). These pumps apply pressure to the float chamber to send fuel up through the main nozzle and into the carburetor throat. Remove the air-filter element and observe if fuel wets the carburetor bore

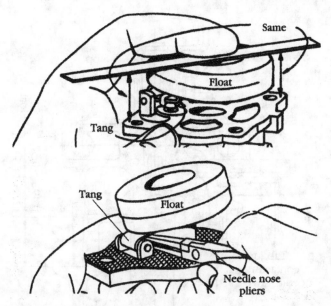

FIGURE 4-15. *Float height is how far the float stands off from the roof of the float chamber. Here the top of the float is the reference point.*

when the bulb is cycled. Malfunctions nearly always are the result of cracks or loss of resiliency in the primer bulb, although the check valves may also fail. Briggs likes to mount primer bulbs on a plastic air-cleaner backing plate, which distorts and leaks pump pressure.

Tecumseh primer bulbs are secured by a plastic star-shaped snap ring. Grasp the bulb with pliers, twist, and pull it out. The snap ring can be retrieved with a dental pick. To install a new bulb and replacement ring, place the ring over the bulb with the tabs up. Then, applying palm pressure to a ¾-in.-deep well socket, force the assembly home.

Briggs primer-bulb retainer rings snap into the air-cleaner backing plate. Depress the locking tabs with a flat-blade screwdriver and extract the bulb. The ball-and-seat valve drops out. To install, moisten the primer bulb, place the locking ring over it, insert the ball and seat, align the two tabs with slots in the housing, and tap the assembly home with a plastic mallet and a 19-mm socket. Briggs PN 19461 makes the job easier, but costs nearly $30.

Manual choke

About the only maintenance required is to see that the choke butterfly pivots open without binding and closes under spring tension. Adjust the

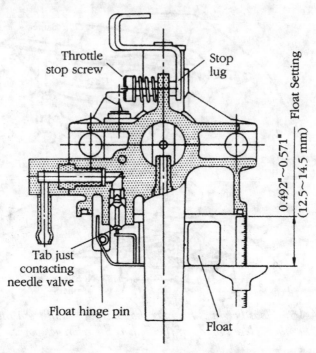

FIGURE 4-16. *Nikki measures float height from the underside of the float. The carburetor should, of course, be inverted when making the measurement.*

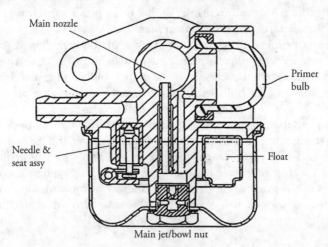

FIGURE 4-17. *Depressing the primer bulb pressurizes the fuel bowl with air to force fuel up through the nozzle and into the carburetor throat. This drawing also depicts the clog-prone Tecumseh and Walbro main jet/bowl nut.*

control cable or linkage so that the choke closes fully when starting and opens fully as the engine warms. A partially open choke will prevent a cold engine from starting. If a warm engine runs better with the choke partially closed, the mixture is lean. Suspect air leaks downstream of the venturi or clogged carburetor jets.

Normally the choke butterfly is left undisturbed when cleaning the carburetor. If for some reason you want to disassemble the unit, note how the cutaway or other markings on the choke plate are oriented; most choke butterflies are unsymmetrical and have beveled edges to make a close fit with the carburetor bore. Metal, as opposed to plastic, choke plates secure with brass screws with their ends peened over. Grind off the peening with a Dremel tool and back out the screws. Brian Miller (1-573-256-0313, pullingtractor@aol.com) is the only source I know of for replacement choke and throttle plate screws.

Automatic choke

An engine's demand for fuel is highest during cold cranking. Once the engine starts, the choke plate should snap partially open and then progressively open wider as the engine warms. In the Kohler design a diaphragm actuator responds to crankcase vacuum by cracking the choke open upon startup (Fig. 4-18). A heat-sensitive bimetallic spring then takes over to open the choke fully within three to five minutes of startup.

Once you have established that the choke plate moves without binding, attach a vacuum pump to the actuator inlet fitting. The choke plate should move half or three-quarters of the way open and remain in that position for at least three seconds. If the plate does not respond and the hose appears sound, replace the diaphragm assembly. The $60 bimetallic spring must be replaced if it fails to open the choke fully after several minutes of running.

Briggs uses a bimetallic spring heated by exhaust temperature as the primary means of opening the choke (Fig. 4-19). An air vane, energized by the flywheel fan, cracks the choke open wider upon startup (Fig. 4-20). Lightly greasing the air-vane pivot eliminates most starting difficulties.

Honda's AutoChoke employs a wax-powered linear motor to open the choke plate against spring tension. The motor, consisting of a wax pellet and a moveable piston, mounts in a recess on the engine block. As the engine warms, the wax melts and expands to drive the linkage that opens the choke. When the engine cools, the wax shrinks back into a solid. Unlike the Briggs and Kohler chokes, there is no mechanism for opening the choke wider when the engine starts before the wax motor fully extends. However, the choke plate is offset, which means that it will tend to pivot open under the force of the air stream.

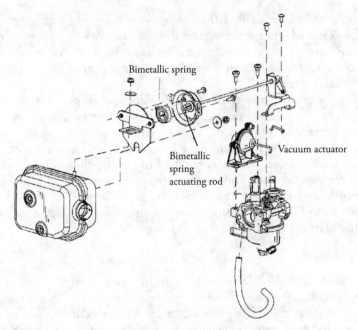

FIGURE 4-18. *Kohler automatic choke.*

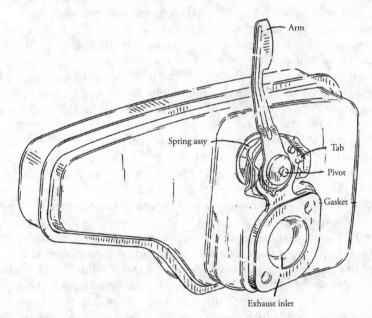

FIGURE 4-19. *The thermostatic spring assembly used on late-model Briggs auto chokes.*

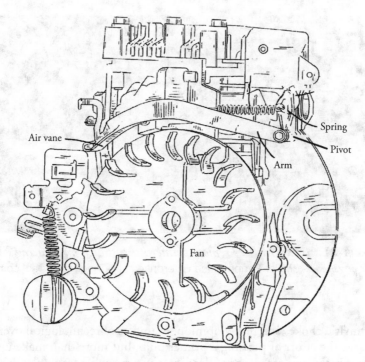

FIGURE 4-20. *Air vane and associated parts. The pivot should be periodically cleaned and greased.*

A lever on the side of the carburetor permits the operator to keep the choke on longer during cold weather and open sooner in summer. Bending a tab on the lever converts the carburetor to a manual choke.

If the engine shuts down before being fully warmed up, the auto-choke may open to make restarting difficult or impossible. Wait for the engine to cool and the choke to close before cranking. Once started, allow the engine to run for at least three minutes before shutting down and restarting.

The mechanism functions nomally if

- The choke plate completely closes on a cold engine.
- The choke plate flutters as it arcs open and remains full open after five to eight minutes of engine operation.

Work the choke lever by hand: the plate should move freely and, upon release, snap back to the closed position. If the linkage checks out okay, replace the wax cylinder.

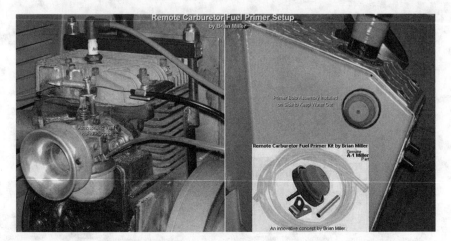

FIGURE 4-21. *Brian Miller's remote primer kit installed on a racing Kohler engine.* A-1 Miller Performance Enterprises at pullingtractor@aol.com

Automatic chokes are, in my opinion, examples of technological overkill. They save the effort of pressing a primer bulb, but more than make up for it in frustration and repair costs. Readers who are having trouble with these devices may want to investigate Brian Miller's primer kit, adapable to any small engine (Fig. 4-21). The kit includes instructions for sealing the original float-chamber vent and a small brass tube that adapts the primer hose to the float-bowl vent (Fig. 4-22).

Diaphragm carburetors

Diaphragm carburetors are used on leaf blowers, chainsaws, string trimmers, and other handheld tools capable of operating at any angle, even inverted (Fig. 4-23). For reasons that are not clear, these rather temperamental carburetors can also be found on some mowers and gensets.

These carburetors maintain their internal fuel level with a diaphragm that balances fuel pressure against atmospheric pressure. The underside of the metering diaphragm opens to the atmosphere; the upper side comes under 5–9 psi of fuel pressure. As fuel is consumed, pressure in the metering chamber drops and the diaphragm, responding to atmospheric pressure on its underside, moves up to unseat the inlet needle. Fuel enters and pressurizes the metering chamber. The diaphragm responds by moving down and out of contact with the inlet needle, which then reseats to block further fuel entry.

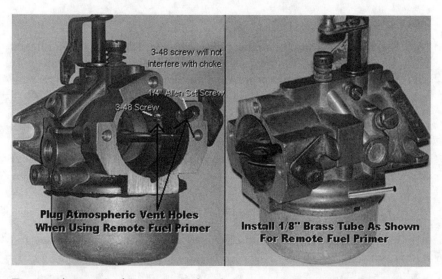

FIGURE 4-22. *Carburetor vent details.* A-1 Miller Performance Enterprises at pullingtractor@aol.com

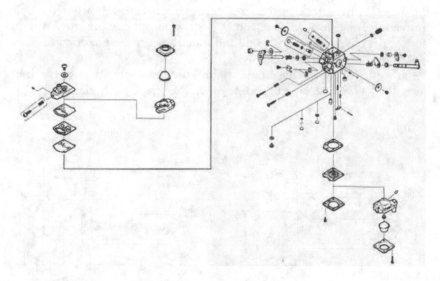

FIGURE 4-23. *Butterfly-throttle diaphragm carburetor.*
Used by permission of Walbro, LLC

Fuel entry

Most diaphragm-carburetor applications include a fuel pump, either integral with the carburetor or mounted remotely. Crankcase vacuum and pressure pulses in two-stroke engines or intake manifold pulses in four strokes, conveyed through a drilled passage in the carburetor body or by an external hose, provide energy (Fig. 4-24). The diaphragm lifts under vacuum pulses to increase the area in the pump chamber. This low-pressure zone draws fuel into the chamber through the inlet check valve. The companion positive pulse drives the diaphragm down to force fuel through the outlet check valve and into the inlet needle and seat.

Fuel delivery

Figures 4-25 through 4-27 illustrate fuel delivery at idle, off-idle, and wide-open throttle.

Nozzle check valve and capillary screen

Most diaphragm carburetors have a check valve that opens under venturi vacuum to enable fuel to discharge through the main nozzle. As shown in Figure 4-25, the check valve closes during idle when the carburetor bore upstream of the throttle plate comes under atmospheric pressure. Were it not for the check valve, air would infiltrate through the nozzle to displace fuel from the idle circuit.

Rather than a check valve, some Walbros use a capillary screen to block air entry (Fig. 4-28). Because of its fine mesh, the screen easily clogs.

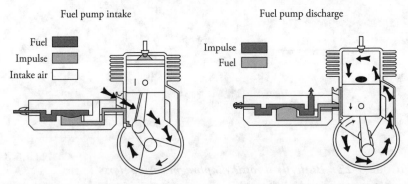

FIGURE 4-24. *Diaphragm pumps operate by vacuum and pressure pulses. A two-stroke engine, shown here, energizes the pump with crankcase pulses. Four-stroke pumps are powered by pressure fluctuations in the intake manifold.* Used by permission of Walbro, LLC

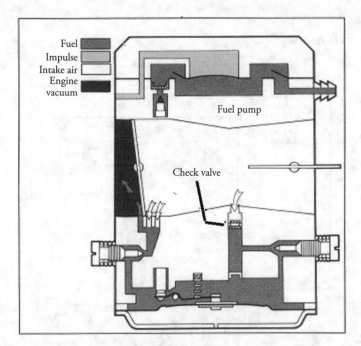

FIGURE 4-25. *Idle operation. At very low speeds only the primary idle port discharges fuel. The secondary, or off-idle, ports aerate fuel waiting for discharge.* Used by permission of Walbro, LLC

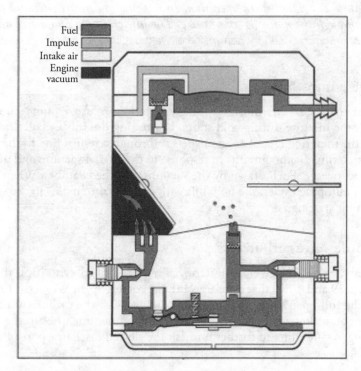

FIGURE 4-26. *Part throttle. The secondary idle ports come on line and the main nozzle check valve opens to permit fuel flow.* Used by permission of Walbro, LLC

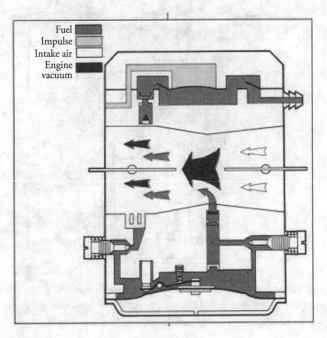

FIGURE 4-27. *High-speed operation. In response to venturi vacuum, fuel flows through the main jet and nozzle. The idle circuit is also active on some carburetors.* Used by permission of Walbro, LLC

Purge pump

Rather than a choke, most diaphragm carburetors use a purge pump to enrich the mixture during cold starts. Depressing the rubber ball draws air from the metering chamber and expels it through a return line to the tank. The metering diaphragm lifts in response to the partial vacuum and unseats the inlet needle. Fuel, drawn by the vacuum, fills the chamber. When, after several pumping cycles, the bulb fills with fuel, no air remains in the system to inhibit starting.

Throttle valve carburetors

A rotary, or barrel-valve, carburetor has a cross-drilled cylindrical throttle (Figs. 4-29 and 4-30) that cams upward as it rotates open. When the throttle is in the full-open position, the cutout merges with the carburetor bore to present an unimpeded path for airflow. At the same time, the attached needle (Fig. 4-30C) lifts to further unmask the slotted discharge port (4-30B).

Cup plug over screen

FIGURE 4-28. *A capillary screen can substitute for a main-nozzle check valve on Walbro carburetors. Remove the cap plug over the screen with a small self-tapping screw. A new screen installs with a pencil eraser.* Used by permission of Walbro, LLC

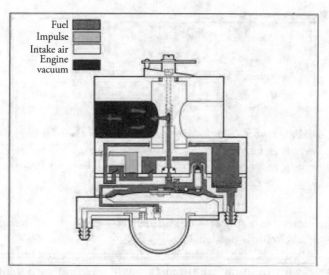

Fuel
Impulse
Intake air
Engine
vacuum

FIGURE 4-29. *A barrel-valve carburetor delivers fuel by way of a retractable needle.* Used by permission of Walbro, LLC

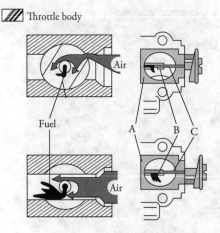

Throttle body

Air

Fuel

A   B   C

Air

FIGURE 4-30. *Barrel-valve carburetor operation: as the throttle (A) turns open, it admits more air and lifts the needle (C) out of the discharge port (B) for increased fuel delivery.* ZAMA Group

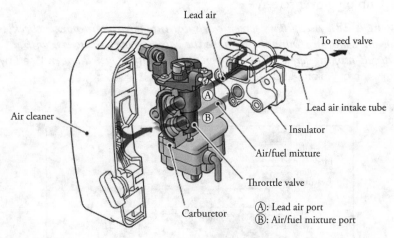

Lead air

To reed valve

Air cleaner

Ⓐ

Ⓑ

Lead air intake tube

Insulator

Air/fuel mixture

Throttle valve

Carburetor

Ⓐ: Lead air port
Ⓑ: Air/fuel mixture port

FIGURE 4-31. *Double-barreled carburetor used on RedMax Strato-Charged engines. The upper barrel supplies air that displaces the fuel charge that would otherwise short circuit out the exhaust port.* RedMax Komatsu Zenoah America, Husqvarna Division

Fuel flow increases in concert with air delivery across the entire rpm band. No low-speed circuit is needed.

As described in Chap. 1, stratified-charge two-strokes buffer the air/fuel charge with fresh air admitted through a second carburetor barrel. Figures 4-31 and 4-32 illustrate the operation of these carburetors.

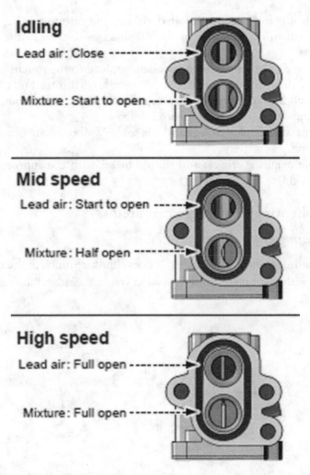

Idling

Lead air: Close

Mixture: Start to open

Mid speed

Lead air: Start to open

Mixture: Half open

High speed

Lead air: Full open

Mixture: Full open

FIGURE 4-32. *The lead air valve initially lags the throttle valve but opens fully at high speed.* RedMax Komatsu Zenoah America, Husqvarna Division

Diaphragm carburetor service

The hundreds of varieties of diaphragm carburetors make a detailed description of servicing impractical. However, there are general points to observe:

- Although fuel and purge pumps may be mounted remotely, most carburetors sandwich these pumps on opposite sides of the metering section. Tighten screws in an X-pattern in order not to distort the castings.

- Repair kits include an exploded view of the carburetor. Pay particular attention to the position of the diaphragms relative to the associated gaskets.
- Replace diaphragms if torn, swollen, or dead to the touch.
- Do not disassemble more than necessary: leave throttle and choke plates as found and be leery of removing Welch plugs. Zama rotary throttles cam upward with an easy-to-lose roller.
- Screens at the inlet fitting and on some Walbro main nozzles require cleaning.
- Needle height is critical and should be set with the appropriate gauge (Fig. 4-33).

Troubleshooting diaphragm carburetors

Pressure testing

While pressure testing is no panacea, it does eliminate much of the uncertainty associated with diaphragm carburetors and crankcase induction. You will need

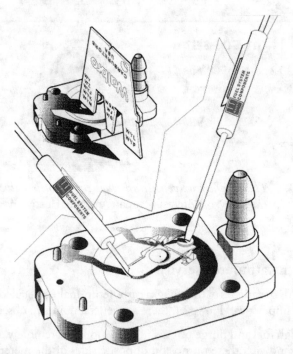

FIGURE 4-33. *Walbro needle height adjustment requires a factory gauge, available from Amazon and other suppliers. Note the use of a small screwdriver to shield the elastomer tip of the needle from bending forces.*

an OEM pressure tester like the one shown back in Figure 4-1 or a Mityvac pressure and vacuum pump.

Begin with the carburetor still installed on the engine. Connect the pump output to the fuel inlet fitting on the carburetor, and apply 10-psi maximum pressure. Wipe up any spilt fuel, disconnect and ground the spark-plug lead, and crank the engine over several times. The gauge should flicker and register reduced pressure as the inlet needle seats and unseats.

Gauge holds steady as the engine is cranked.	Inlet needle not unseating. No fuel can enter the instrument.	Gummed needle, malfunctioning metering diaphragm or (which is less likely) a failed pump discharge check valve.
		Swollen shut or leaking fuel line.
Pressure drops without cranking	Air leaks—engine runs lean or not at all. *or* Inlet needle not seating. Carburetor floods.	Defective metering chamber or fuel-pump cover gaskets. *or* Inlet needle stuck open; needle and/or seat worn or contaminated.

If you detect a pressure leak, immerse the carburetor in mineral spirits or water with the tester hose connected to the inlet fitting as before. Apply no more than 10 psi to the fitting and watch for bubbles. Leaks around the diaphragm cover mean that the gasket has failed or that the cover is warped. The same holds for the fuel-pump cover, although a tiny air leak in this area does not always translate as a fuel leak. A leaking inlet needle emits air bubbles from the main discharge nozzle. Attempt to repair the leak by replacing the diaphragm, needle, and spring. Should the leak persist, the needle seat is at fault. With the exception of Tillotson carburetors, current diaphragm carburetors do not have replaceable needle seats. You will have to replace the carburetor.

Most inlet needles pop off, or unseat, at between 9 and 12 psi. Some may take higher pressures, and seemingly identical carburetors exhibit markedly different pop-off pressures. Reseat pressures should be 10 psi or higher. Otherwise, the needle will not seat against pump pressures that typically range between 5 and 7 psi.

For accurate readings, the needle must be wetted with mineral spirits or WD-40. Connect the tester to the carburetor inlet fitting and slowly build pressure. Pressure will suddenly drop as the needle pops off and then stabilize at a lower pressure. This hopefully steady pressure is the reseat pressure. Carburetor gurus change these pressures by tinkering with spring tension. For the rest of us, it is prudent to let the factory setting stand.

There are times when nothing that one does coaxes sufficient fuel from a carburetor. The spark-plug tip will be bleached white or gray, and the engine will have to be chocked to develop anything like normal power. A certain percentage of carburetors are not rebuildable. Boil 'em, dip 'em in acid, scrub 'em an ultrasonic cleaner, and they still refuse to deliver enough fuel. But the carburetor may not be at fault.

Two-stroke engines are an extended induction track. Air leaks around the main-bearing seals or across any gasketed surface cost vacuum and dilute the fuel charge. To pressure test an engine, it is necessary to remove the shrouding and fabricate cover plates for the intake and exhaust ports (Fig. 4-34). One of the plates has an inlet fitting that mates up to the tester hose. Plate gaskets can be cut from a bicycle tube.

Lower the piston to bottom dead center. You can use a screwdriver inserted through the spark-plug hole to indicate piston position. Replace the spark plug, connect the hand pump, and pressurize the cylinder to 6 psi. Higher pressures risk blowing out the crankshaft seals.

FIGURE 4-34. *Cover plates for a weed-trimmer engine.* Photo by Tony Shelby

While leakage should ideally be no more than 1 psi per hour, the loss of 1 psi per minute is worrisome, but tolerable. Soapy water locates the leak sources that are most likely to be at the main bearing seals. Other potential leak sources are the

- Crankcase cover gasket
- Carburetor pulse port
- Reed valve cage
- Base gasket on engines with detachable cylinder jugs
- Spark-plug gasket

Crankcase leaks can also originate at blowholes in the casting, although this sort of defect is rare.

Diaphragm carburetor malfunctions

No or hard starting

Refusal to start can almost always be attributed to insufficient fuel delivery. The spark-plug tip will remain dry after repeated cranking. Verify by spraying carburetor cleaner into the cylinder at the spark-plug port. If the engine runs a few seconds, we can be sure that the problem is fuel starvation.

Warning: Carburetor or brake cleaner aerosols are both toxic and extremely flammable.

Possible causes of hard starting:

- Air leaks at the fuel-line connections.
- Air leaks at the fuel-pump pulse line (if fitted) or a clogged crankcase pulse port on two-stroke engines. To verify that the pulse port is clear, remove the carburetor and apply a few drops of oil to the port. Crank the engine through. The oil should bubble up in response to crankcase pressure. If not, the port is carbon clogged.
- Fuel-pump failure involving the cover gasket, diaphragm, inlet screen, or check valves.
- Metering system malfunctions, which can be traced to a stuck needle, unresponsive diaphragm, leaking cover gasket, or a clogged discharge nozzle screen.
- Purge pump failure or a choke plate that hangs open.

Flooding

Flooding is caused by:

- Dirt or gum that prevents the inlet needle from seating.
- Improperly installed metering diaphragm.
- Diaphragm spring not seated on its dimple.

Refusal to idle

Failure to idle can have several causes:

- Improperly adjusted idle stop screw.
- Improperly adjusted low-speed mixture screw (when present).
- Diaphragm-lever spring not seated.
- Clogged idle circuit.
- Air leaks around the throttle shaft on high-hour engines, a loose carburetor flange, or crankcase air leaks.
- Leaking main nozzle check valve that permits excess air to enter the idle circuit at low engine speeds. Test the check valve by blowing on a tube cut square and pressed against the fuel entry passage at the base of the nozzle. The valve should open under pressure and close when you inhale.
- Welch plug covering idle circuit does not seal.

Erratic idle

Adjust the low-speed mixture screw (if present) to obtain the smoothest idle. Four-stroking—duh, duh, POW! duh—can often be attributed to fuel puddling in the crankcase. At least we hope so, since that malfunction is a design problem for which mechanics are blameless.

Ragged acceleration

Bogs or stumbles as the throttle is snapped open mean that the mixture has gone lean.

- Low-speed mixture screw (when present) adjusted too lean
- High-speed mixture screw (when present) adjusted too lean. Turn counter-clockwise to enrichen
- Metering lever set too low
- Loose metering diaphragm cover plate
- Leaking metering chamber gasket
- Ruptured or unresponsive fuel-pump diaphragm
- Main nozzle orifice clogged
- Faulty fuel-tank vent

Lack of power

Malfunctions are similar to, but less pronounced than, those listed under the "No or hard starting" heading earlier.

Suction-lift carburetors

Carburetors that mount on top of the fuel tank and draw through a tube were once popular, but now are confined to a few low-end Briggs & Stratton engines. I'm not going to go into much detail, since readers who have become acquainted with diaphragm carburetors have nothing to fear from devices that work like flit guns.

Briggs suction-lift carburetors underwent two basic design evolutions. The earlier versions, known as Vacu-Jets, have a single pickup tube and plug-type choke (Fig. 4-35). They worked, but barely. As the tank emptied, the fuel level in the pickup tube dropped, and the mixture leaned out.

Briggs addressed the problem with the Pulsa-Jet. This carburetor employs two pickup tubes and an integral pump to move fuel from the tank to a small reservoir. The carburetor draws fuel from the reservoir, which is continuously topped off by the diaphragm pump. Thus, the level of fuel in the tank has no effect on the mixture. Depending upon the model, the pump diaphragm mounts on the side of the carburetor or between the carburetor and tank (Fig. 4-36).

The newest carburetor in the series is the plastic Pulsa Prime currently available as the PN 795475 replacement for the original PN 790206. Because these carburetors can be had for as little as $16, few shops bother with repairs.

Suction-lift carburetor service

The usual malfunctions involve failure to deliver fuel caused by a stretched pump diaphragm (Pulsa-Jet) or a sticking check valve (Vacu-Jet). Free the check ball with a piece of thin wire inserted through the pickup tube screen. Low-speed ports on both carburetors are under the mixture-adjustment screw and can be cleared with carb cleaner or compressed air.

Pulsa- and later-production Vacu-Jet carburetors usually come with vacuum-operated chokes. A spring-loaded diaphragm holds the choke closed during cranking. Once the engine starts, manifold vacuum acts on the underside of the diaphragm to pull the choke open. The choke also serves as an enrichment valve: when the engine slows under load, the loss of manifold vacuum causes the choke to close.

Assembling the choke diaphragm is a bit tricky:

1. Clip the choke link to the diaphragm as shown in Figure 4-37.
2. Invert the assembly and snake the linkage into the recess (Fig. 4-38). Running two of the mounting screws into their holes on the carburetor mounting flange will keep the diaphragm in alignment as the carburetor is mated up to the tank. Thread in the mounting screws lightly.
3. Hold the choke butterfly closed with one finger, and attach the connecting link to the choke (Fig. 4-39).
4. Tighten the mounting screws in an X-pattern. Diaphragm preload should hold the choke plate lightly closed.

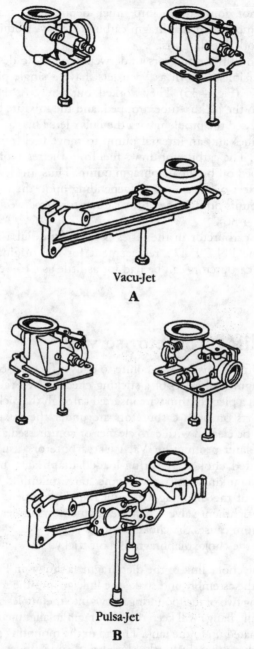

Vacu-Jet

A

Pulsa-Jet

B

FIGURE 4-35. *Briggs & Stratton Vacu-Jet (A) and Pulsa-Jet (B) carburetors.*

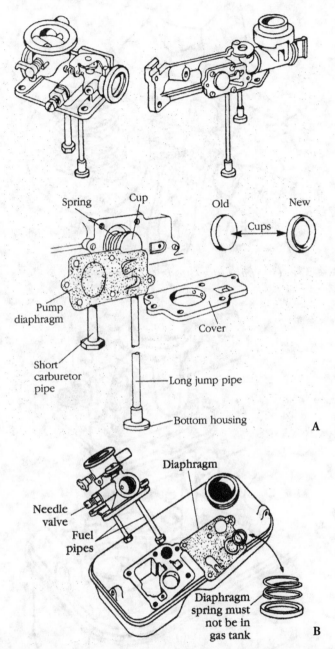

FIGURE 4-36. *Pulsa-Jet malfunctions nearly always involve the diaphragm, which mounts on the side of the carburetor body (A) or between the carburetor and tank (B). Mating surfaces on the tank must be dead flat to make an airtight seal.*

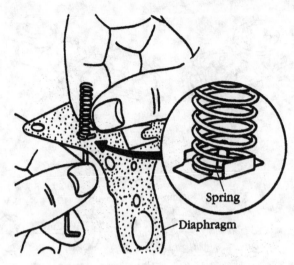

FIGURE 4-37. *The choke link clips to the diaphragm.*

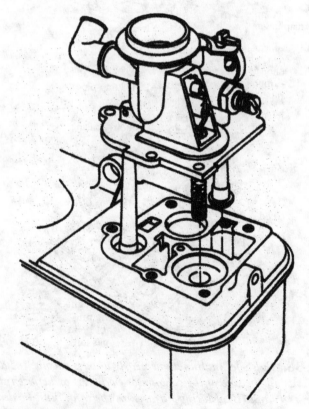

FIGURE 4-38. *Lightly engage the mounting screws in their threads. The carburetor and tank should be free to shift about.*

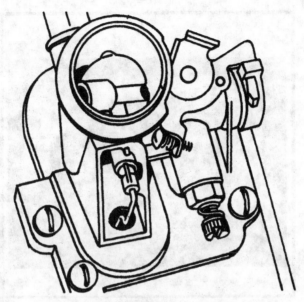

FIGURE 4-39. *To establish the preload, it is necessary to connect the linkage while holding the choke butterfly closed. When installed correctly, the choke will close during engine starts, flutter under acceleration and load, and open when engine speed holds steady. Test with the air-filter screw installed.*

Carburetor adjustments

The classic carburetor has three adjustment screws—idle rpm, low-speed, or idle mixture, and high-speed mixture. The idle-rpm screw merely functions as a throttle stop and has no effect upon mixture strength. The low-speed mixture screw threads into the carburetor body near the throttle blade. Backing the screw out richens the mixture. The high-speed mixture screw, also known as the main adjustment needle or the main fuel needle, works in conjunction with the main jet to regulate mixture strength at wide throttle angles. It may be located under the float bowl or on the carburetor body upstream of the idle-mixture screw.

Suction-lift and barrel-valve diaphragm carburetors have either a single screw or fixed jets.

California Air Resources Board (CARB) and the Environmental Protection Agency (EPA) mandate that the ability to change factory mixture adjustments be confined to certified technicians. Idle rpm is the only adjustment that consumers are trusted to make. Hidden adjustment screws, travel limits, "tunnel enclosures," and propriety screw heads make the adjustments difficult, but not impossible for those willing to void factory warranties and

FIGURE 4-40. *Adjustment needles buried in "tunnels."* California Air Resources Board

risk the wrath of regulatory agencies (Figs. 4-40 and 4-41). The Internet explains how.

Seat adjustment screws gently: forcing them home damages the needles and their seats (Fig. 4-42). Backing out the screws one or two turns should be enough to get the engine started.

Final adjustments are made with the engine warmed up and the tank half-full of fresh gasoline. Suction-lift carburetors with a single mixture screw are adjusted for best idle. Snap the throttle open with your finger. If the engine hesitates, open the screw further and retest.

To adjust carburetors with idle- and high-speed mixture screws:

1. Run four-stroke engines at about three-quarters throttle and two-stroke engines at near full throttle.
2. Back out the high-speed screw in small increments of one-eighth of a turn. Pause for a few seconds after each increment to permit the engine to respond. Stop when the engine speed falters at the rich limit. Note the position of the screw slot.
3. Tighten the high-speed screw in the same one-eighth-turn increments as before. Stop when the engine rpm drops in what is aptly called "lean roll." Note the position of the screw slot.

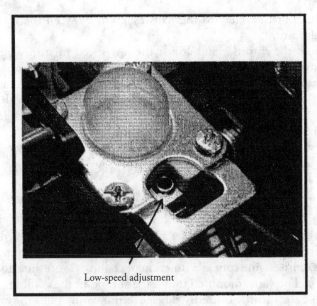

Low-speed adjustment

FIGURE 4-41. *Adjustment screws can be hidden under expansion plugs or plastic caps.* California Air Resources Board

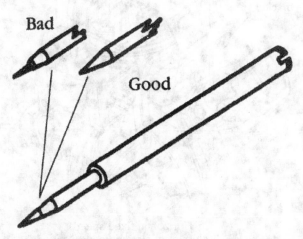

FIGURE 4-42. *Bent needles or needles with wear grooves deep enough to be felt need replacement. The undercut needles supplied with some Walbro carburetors shear off when overtightened.*

4. Back out the screw to the halfway point between the lean roll and rich limit.
5. Close the throttle, and adjust the low-speed mixture screw for the fastest idle. If the engine overspeeds, back out the idle stop screw.
6. Snap the throttle open with your finger. If the engine hesitates, richen the mixture. Some carburetors accelerate more smoothly with a slightly rich low-speed mixture; others respond to a richer high-speed mixture.
7. Test the engine under load. The high-speed mixture may need to pass more fuel at the expense of a slightly soggy idle.

Governors

A unique feature of utility engines is the governor mechanism that maintains engine speed under load and limits wide-open-throttle rpm. Air-vane governors sense engine rpm as a function of cooling air pressure and velocity (Fig. 4-43). The vane, mounted under the shroud in the cooling air stream, acts to close the throttle butterfly. The governor spring pulls the butterfly in

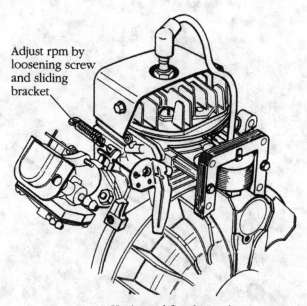

Adjust rpm by
loosening screw
and sliding
bracket

Horizontal fixed speed
(aluminum air-vane governor)

FIGURE 4-43. *Air-vane governor used on Tecumseh two-strokes. Like some Kohler engines, rpm is set by a sliding bracket.*

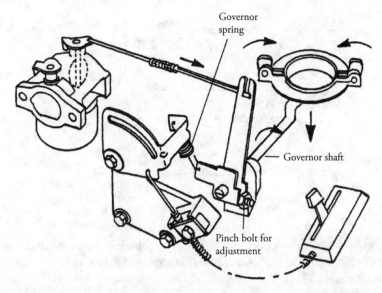

Governor
spring

Governor shaft

Pinch bolt for
adjustment

FIGURE 4-44. *Mechanical governors use centrifugal force acting on a pair of spinning weights to close the throttle against spring tension. As the engine slows under load, the force exerted by the weights diminishes and the spring opens the throttle butterfly.*

the opposite direction. If speed drops below a certain value, the spring pulls the throttle open.

Mechanical governors sense engine speed with centrifugal flyweights (Fig. 4-44). Centrifugal force pivots the flyweights outward. This movement, translated through a spool and yoke, applies more force to close the throttle as engine speed increases. The governor spring opposes this force. As speed drops, the force generated by the flyweights diminishes and the spring pulls the throttle butterfly open.

More elaborate governor mechanisms incorporate a low-speed adjustment (distinct from the throttle butterfly stop screw) and have a provision to adjust governor sensitivity, or speed droop. Maximum rpm can be varied by moving the governor arm relative to the shaft, but no universal procedure applies (Fig. 4-45). Governor adjustments should be left to the dealer.

Electronic fuel injection

Emissions from small engines are difficult to control since rich air/fuel mixtures are required to provide good load pickup, speed stability, and, on air-cooled engines, cylinder cooling. Carburetors regulate fuel delivery by

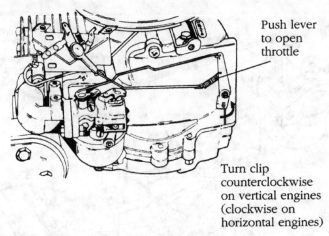

Push lever
to open
throttle

Turn clip
counterclockwise
on vertical engines
(clockwise on
horizontal engines)

FIGURE 4-45. *The relationship between the governor shaft and wide-open throttle is a crucial aspect of small-engine service. Make a mistake and the engine grenades. As indicated in the drawing, adjustment procedures are not standardized.*

means of manifold vacuum, but vacuum says nothing about engine temperature, ambient air pressure, or the dozen other variables that affect engine performance. Nor do carburetors respond well to transient conditions. The mixture goes lean as the throttle snaps open and goes rich when the carburetor idles down. Provisions for cold starting are at best crude.

Electronic fuel injection (EFI) enhances responsiveness and reduces hydrocarbon emissions by 40 percent and carbon monoxide by 80 percent, but with some increase in oxides of nitrogen due to higher combustion temperatures. The major benefit, which enables commercial owners to pay for the technology, is a 20 percent or better improvement in fuel economy. On the other hand, replacement parts tend to be expensive and labor charges often reflect the inexperience of dealer mechanics. EFI is currently available on Briggs & Stratton Vanguard and on top-of-the-line Kawasaki, Subaru, Onan and Kohler engines.

Documentation and tools

EFI systems are relatively simple in concept, but test procedures and trouble codes vary by make. Having access to the appropriate shop manual saves time and frustration.

- Briggs & Stratton—available for purchase from local dealers or https://shop.briggsandstratton.com/us/en/parts-and-accessories/repair-manuals

- Kawasaki—available for purchase from local dealers
- Kohler—free, downloadable at https://www.kohler-engine-parts
.opeengines.com/index.php?main_page=document_general_info
&products_id=34596
- Onan—available for purchase from Cummins Power Generation
dealers
- Subaru—free, downloadable at http://www.subarupower.com/
products/manuals/

The primary tool is a digital volt-ohm-meter (DVOM) with at least a 10-megaohm input impedance. A test lamp is often more convenient to use, but must not draw more than 0.3 A (ampere). Design Technology sells $100 Briggs and Kohler scan tools and back probe test leads for checking voltage on made-up connections. You can also fashion a back probe from a straightened paper clip, which is then carefully inserted into the back of the connection to make contact with the desired pin. A node lamp can be purchased from the engine manufacturer for checking that injectors receive power. Auto parts stores also sell node lamps, but finding one with the correct terminal geometry can be difficult.

Precautions

- Most of these systems operate at 40 ± 3 psi of pressure, and all of them hold pressure after the engine is stopped. Wait until the engine is cold before opening a fuel line, wear eye protection, and have rags ready to wipe up the spills. Replacement injector and other high-pressure lines, together with Oetiker hose clamps, are best purchased from the engine manufacturer. If OEM (original equipment manufacturer) replacement hoses are not available, use SAE J30R9 hose designed for fuel-injection systems.
- Disconnect both battery terminals, and remove the battery from the equipment when charging the battery.
- Do not start the engine with loose battery terminals or disconnect a battery terminal while the engine is running.
- Do not disconnect the alternator lead from the battery or the battery from the EFI harness while the engine is running.
- Do not connect an external voltage source to any EFI component.
- Do not cycle injectors or power up an electric injector pump without fuel. Gasoline is an essential lubricant.
- Do not open any wiring harness connector with the key switch on or with the engine running. To do so is to risk frying the electronic control unit (ECU).

Troubleshooting

Most malfunctions—no start, lack of power, misfiring—are engine related. Check the ignition, compression, and crankcase vacuum. Spark plugs are a prime suspect. Malfunctions that involve the fuel system usually are the result of dirty filters, a plugged fuel-tank cap vent, or air leaks in the plumbing or induction tract. It's good practice to vacuum test fuel-line integrity. If the spark plugs remain dry after repeated cranking, check for the presence of fuel at each component from the tank to the injector supply lines. As noted previously, wear eye protection and have a supply of rags ready to wipe up the spills. A spark, as from a dropped tool or a discharge of static electricity accumulated as we move about, can have horrific consequences in the presence of spilled fuel.

Electrical problems most often involve the battery or wiring harness or the most obvious of all—fuses. An EFI requires more than 8 V of cranking voltage to function. Inspect the harness for corroded connections; chaffed insulation; and rust, paint, or grease on the chassis ground connections. Ignition leads should be routed as far as possible from EFI wiring.

Voltages

Two voltages—12 V and 5 V $\pm$ 0.1 V—are present. The former supplies power to the fuel pump and, on engines so equipped, to the throttle motor and oxygen-sensor heater. Measure the voltage while the engine runs by back-probing the connectors with a DVOM and factory-supplied back-probe test leads. If you lack the factory tool, use a paper clip. The other DVOM lead goes to a good (paint-, rust-, and grease-free) engine ground.

The ECU reports sensor malfunctions as abnormal variations in voltage. Sensors operate like variable resistors. How much resistance the sensor develops appears as the signal voltage, which can range from 0.1 to 04.9 V. Abnormalities arise from:

- Low, high, or no reference voltage
- Low, high, or no signal voltage

High reference voltages are hardly even encountered and result from a failed voltage-regulation circuitry in the Engine Control Module (ECM). Low or no reference voltage usually can be traced to a connector with corroded, dirty, or bent pins. Sometimes wiggling the connector will reveal the fault. Abraded or pinched wires result in shorts to ground.

When measured at the sensor connection, low or no signal voltage means that the sensor has malfunctioned. No sensor voltage at the ECM harness connection indicates a wiring harness problem. Use the wiring color code to keep track of the pin connections.

Sooner or later one encounters a sensor that lies, but not so obviously that it flags a trouble code. For example, an engine temperature sensor may report 20°F, when the actual block temperature is 150°F. The computer richens the mixture as if the engine had just started on a cold morning. Bosch and Delphi publish graphs that correlate sensor resistance with temperature, but applying that data in the field verges on the impossible. One ends by throwing parts at the problem.

EFI operation

Small-engine EFIs are scaled-down and somewhat simplified versions of automotive systems, designed to easily replace carburetors on existing engine models (Table 4-1).

ECU

The electronic control unit (ECU) (or, more simply, the computer) integrates sensor data to calculate the pulse width, or how many milliseconds the injectors remain open. The most basic EFI, known as the Alpha-N system, operates in the context of engine rpm, throttle angle, and, for cold starting, engine temperature. The faster the engine turns and the greater the throttle angle, the longer the injectors remain open. While Alpha-N systems deliver less-than-optimal fuel economy, they have the advantage of simplicity. The designers of Subari's Alpha-N EFI did not feel it necessary to provide fault codes: visual inspection and some deductive skills were considered adequate for troubleshooting. Since the ECU remembers no history, a battery was not needed, which further simplies matters.

Briggs, Kohler, Kawasaki, and Onan opted for the more sophisticated Speed-Density system that employs a manifold absolute pressure (MAP) sensor to signal engine load. When the key switch is turned to the On position, the MAP sensor reports barometric pressure to the ECU, which uses this data to calculate air density and, indirectly, the amount of oxygen available for combustion. Once the engine starts, manifold pressure variations reflect engine loads. At idle, the engine produces only enough power to overcome compression resistance and internal friction. The nearly closed throttle blade generates high manifold vacuum. Even higher levels of vacuum are developed during coastdown, when the engine pulls against a closed throttle. The ECU reduces fuel delivery during periods of high manifold vacuum. As loads are encountered, the throttle swings open, manifold vacuum diminishes, and the ECU orders the injectors to remain open longer. In the newer versions of the technology, the sudden pressure loss when an intake valve opens enables the ECU to time injection with piston intake strokes.

Briggs, Kawasaki, and Kohler have a crankshaft position sensor (CPS) to synchronize injection with engine speed and the intake piston stroke.

Table 4-1. EFI system components

Component	Briggs & Stratton	Kawasaki	Kohler	Onan	Subaru
Throttle position sensor	√	√	√	√	√
Manifold absolute pressure sensor	√	√	√	√	
Engine temperature sensor	√	√	√	√	√
MAT sensor	√		√	√	
Crankshaft position sensor	√	√	√	ECU monitors ignition pulses	
Pulse-type transfer pump	√ on Model 49000; electric transfer pump on Model 610000		√		
Injector pump type & location	12 V fuel module	12 V inline	12 V fuel module	12 V inline	Pulse-actuated throttle body
Oxygen sensor	√		√		
Malfunction indicator lamp (MIL)	√	√	√	√	

Kohler and Briggs engines feature closed-loop operation. An oxygen, or O_2, sensor reports the oxygen content of the exhaust gases to enable the ECU to make minor adjustments to fuel delivery.

Throttle body

The throttle body contains the throttle mechanism and a variety of other components. Kawasaki and Onan gen-set throttle bodies include a governor motor, and Kawasaki throttle bodies include the injectors. Briggs throttle bodies have

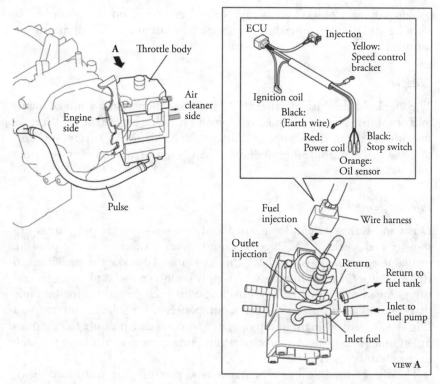

FIGURE 4-46. *Subaru combines an injector pump and pressure regulator with the throttle body.*

an idle air-bleed for stable low-speed operation, and Suzuki (Fig. 4-46) uses the throttle body as a platform for the injector pump and pressure regulator, as indicated by the presence of a return line to the fuel tank. As for the throttle body itself, the only service procedures are to verify that the throttle does not bind and to periodically remove carbon deposits. This last operation can be a bit of a chore on plastic throttle bodies that do not tolerate aerosol carburetor cleaners.

TPS

The throttle position sensor (TPS) mounts on the throttle body concentric with the throttle-blade shaft and uses a rare-earth magnet to induce an increase in signal voltage as the throttle is opened. Verify that it receives a 5 V reference voltage. The signal voltage should be between 0.6 and 1.2 V at idle and increase without skips or hesitation to 4.3 or 4.8 V at wide-open throttle.

Note: If the Kohler TPS is removed or disturbed on its mounting, it must be initialized, a procedure best done with the aid of Kohler diagnostic software.

MAP sensor

The manifold absolute pressure (MAP) sensor mounts on the intake manifold or throttle body. The signal voltage on a running engine should decrease smoothly as the throttle is opened. Briggs and Kohler MAP sensors are combined with a manifold air temperature (MAT) sensor and fed from the same four-wire connector.

Injector pump

Onan and Kawasaki employ gravity feed or a low-pressure pulse pump to transfer fuel to the 12 V injector pump. A return line from the pressure regulator cycles surplus fuel back into the tank. To make the transition to EFI easier for OEM customers who supply their own fuel tanks, Kohler and Briggs opted to eliminate the return line with a self-contained fuel module that contains a high-pressure electric pump, pressure regulator, and reservoir (Fig. 4-47). A diaphragm pump operated by crankcase pressure fluctuations or, on some Briggs engines, an electric lift pump, transfers fuel from the tank to the module reservoir.

The Briggs module has a Schrader-type pressure-test and bleed-down connection. Pump output should be between 35 and 43 psi for Briggs and 37 psi or so for the Kohler unit.

Injectors

When the engine is cranked, the clicks as injectors open and shut can be detected with a stethoscope. A node lamp with the proper connections will reveal if the injectors receive power. Testing injector spray patterns on the engine is a way to join the heavenly choir and is not recommended.

Temperature sensors

The engine temperature sensor is a heat-sensitive variable resistor, or thermistor, which mounts on the cylinder head or block to notify the ECU that the engine is cold and additional fuel is needed. As engine temperature rises, the resistance of the sensor decreases with a corresponding increase in signal voltage. A MAT sensor enables the ECU to better calculate air density. Like the engine temperature sensor, the MAT sensor has a negative temperature coefficient which means that signal voltage increases with heat.

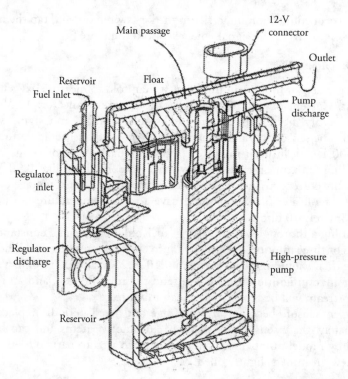

FIGURE 4-47. *The Kohler fuel module consists of a reservoir, a 12 V high-pressure pump, pressure regulator, and float valve. Fuel enters through a tube at the top of the unit under low pressure generated by a vacuum-operated lift pump. When the reservoir is full, the float valve closes to seal the vent tube (not shown in this cutaway view). The diaphragm pump then dead heads to deliver no fuel. As fuel is consumed, reservoir pressure drops and the diaphragm pump resumes operation. The high-pressure injector pump runs continuously. Should output pressure exceed 39 psi, the pressure regulator opens to bypass fuel to the reservoir. The Briggs fuel module is nearly identical.*

CPS

The crankshaft position sensor (CPS) works with a permanent magnet and a toothed reluctor wheel with one tooth missing. The missing tooth upsets the magnetic field to produce an intermittent AC signal once every revolution, which the computer reads as crankshaft position and engine speed. For reliable performance, the magnet should be within 0.30 in. of the target wheel. In theory, you can test a CPS by unbolting it from the engine with the electrical connection intact. Turn the key switch on, and tap the magnet with a screwdriver. The signal voltage should fluctuate.

Oxygen sensor

An oxygen sensor mounts in the exhaust manifold to measure oxygen in the exhaust. Current oxygen sensors have a four-wire connector: 12 V heater power supply, heater circuit ground, signal voltage, and signal voltage ground. The sensor acts as an open circuit when exhaust temperature is less than 600°F. A lean-limit mixture creates a signal voltage of about 0.12 V; a rich-limit mixture increases the voltage to something like 0.90 V. The chemically correct air/fuel ratio of 14.7:1 produces a signal voltage of about 0.45 V. The heater filament should have a room-temperature resistance of approximately 10 ohms.

Running a thoroughly warm Briggs or Kohler engine at a constant speed for two or three minutes puts the EFI into closed-loop operation. The signal voltage then should oscillate between 0.10–0.25 V and 0.70–0.90 V.

Note that an ignition misfire will send the mixture rich, and an exhaust leak upstream will be reported as a lean mixture.

Oxygen sensors become lazy with age and need periodic replacement. Fortunately, these sensors are standard automotive items and can be purchased by model number at a reasonable price at auto parts outlets. Install with new grommets, lightly oiled.

5

Rewind starters

Unlike other engine systems that operate continuously, manual and electric starters are designed for intermittent use, which is why rewind starters can get by with nylon bushings, and why motor pinions can cheerfully bang into engagement with the flywheel. The starter usually lasts about as long as the engine and the owner is satisfied.

But the balance between starter and engine life goes awry if the engine is allowed to remain chronically out of tune. Most starter failures are the result of overuse: The starter literally works itself to death cranking a bulky engine. Whenever you repair a starter, you must also—if the repair is to be permanent—correct whatever it is that makes the engine reluctant to start in the first place.

Side pull

The side-pull rewind (recoil, self-winding, or retractable) starter was introduced by Jacobsen in 1928 and has changed little in the interim. These basic components are always present:

- Pressed steel or aluminum housing, which contains the starter and positions it relative to the flywheel.
- Recoil spring, one end of which is anchored to the housing, the other to the sheave.
- Nylon starter rope, which is anchored to and wound around the sheave.
- Sheave, or pulley.

- Sheave bushing between sheave and housing or (on vintage Briggs & Stratton) between sheave and crankshaft.
- Clutch assembly.

Troubleshooting

Most failures have painfully obvious causes, but it might be useful to have an idea of what you are getting into before the unit is disassembled.

Broken rope is the most common failure, often the result of putting excessive tension on the rope near the end of its stroke or by pulling the rope at an angle to the housing. The problem is exacerbated by a worn rope bushing (the guide tube, at the point where the rope exits the housing). In general, rope replacement means complete starter disassembly, although some designs allow replacement with the sheave still assembled to the housing.

Refusal of the rope to retract. If the whole length of the rope extends out of the housing, either the spring has broken or the anchored end slipped. If the rope retracts part of the way and leaves the handle dangling, the problem is loss of spring preload. The best recourse is to replace the spring, although preload tension can be increased by one sheave revolution. When this malfunction occurs on a recently repaired unit, check starter housing/flywheel alignment, spring preload tension, and replacement rope length and diameter.

Failure to engage the flywheel is a clutch problem, caused by a worn or distorted brake spring, a loose retainer screw, or oil on clutch friction surfaces. While recoil springs and sheave bushings require some lubrication, starter clutch mechanisms must, as a rule, be assembled dry.

Excessive drag on the rope often results from misalignment between the starter assembly and flywheel. If repositioning the starter does not help, remove the unit, turn the engine over by hand to verify that it is free, and check starter action. The problem might involve a dry sheave bushing.

Noise from the starter as the engine runs should prompt you to check the starter housing and flywheel alignment. On Briggs & Stratton in-house designs, the problem is often caused by a dry sheave bushing (located between the starter clutch and crankshaft). Remove the blower housing and apply a few drips of oil to the crankshaft end.

Overview of service procedures

Rewind starters are special technology, and an overall view of the subject is helpful. The first order of business is to release spring preload tension, which can be done in two ways. Any rewind starter can be disarmed by removing the rope handle and allowing the sheave to unwind in a controlled fashion. Other starters have provisions for tension release with the handle still attached to the rope. Briggs & Stratton provides clearance between sheave diameter and housing that allows several inches of rope to be fished out of the sheave groove. This increases the effective length of the rope, enabling the sheave and attached spring to unwind. Other designs incorporate a notch in the sheave for the same purpose (Fig. 5-1).

Brake the sheave with your thumbs as it unwinds. Count sheave rotations from the point of full rope retraction so that the same preload can be applied upon assembly.

The sheave is secured at its edges by crimped tabs and located by the crankshaft extension (Briggs & Stratton side pull), or else it rotates on a pin attached to the starter housing. A screw (Eaton) or retainer ring (Fairbanks-Morse and several foreign makes) secures the sheave to the post.

The mainspring lives under the sheave, coiled between sheave and housing; with its inner, or movable, end secured to the sheave hub. The outer, or stationary, spring end anchors to the housing. Unless the spring is broken, do not disturb it.

Warning: Even after preload tension is dissipated, rewind springs store energy that can erupt when the sheave is disengaged from the housing. Wear safety glasses.

The manner in which recoil springs secure to the housing varies among makes, and this affects service procedures. Many use an integral spring retainer that indexes to slots in the housing (Fig. 5-2). The spring and retainer are serviced as a unit and should not be separated.

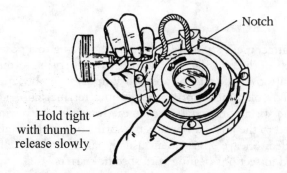

Notch

Hold tight
with thumb—
release slowly

FIGURE 5-1. *Common sense dictates that the starter should be disarmed before the sheave is detached. Most have provision to unwind the rope a turn or so while others are disarmed by removing the rope handle and allowing the rope to fully retract.*

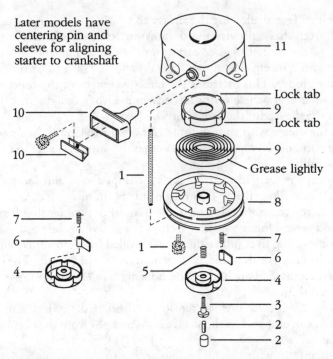

Later models have
centering pin and
sleeve for aligning
starter to crankshaft

11

Lock tab
9
Lock tab
10
9
Grease lightly
10
1
8
7
6
7
1
6
4
5
4
3
2
2

FIGURE 5-2. *Eaton rewind starter, with integral mainspring and housing, should not be dismantled in the field. Lock tabs on the spring-housing rim mean that the spring and housing should not be dismantled. The starter also uses a small coil spring—shown directly below the sheave—to generate friction on the clutch assembly.*

Another attachment strategy is to secure the spring to a post pressed into the underside of the housing. The fixed end of the spring forms an eyelet or hook that slips over the anchor post. To simplify assembly, most manufacturers supply replacement springs coiled in a retainer clip. The mechanic positions the spring and retainer over the housing cavity with the spring eyelet aligned to the post and presses the spring out of the retainer, which is then discarded. Sheave engagement usually takes care of itself. Exceptions are discussed in sections dealing with specific starters.

Some starters adapt to left- or right-hand rotation by reversing the spring (Fig. 5-3). Viewing the starter housing from the underside and using the movable spring end as reference, clockwise engine rotation requires counterclockwise spring windup. The wrap of the rope provides appropriate sheave rotation.

Clockwise engine rotation

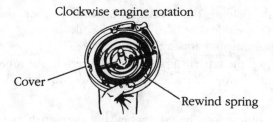

Cover

Rewind spring

Counterclockwise engine rotation

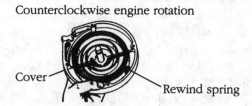

Cover

Rewind spring

FIGURE 5-3. *Many rewind springs and all ropes can be assembled for left or right hand engine rotation. This feature is a manufacturing convenience that makes life difficult for mechanics.*

The spring anchor for traditional Briggs & Stratton starters takes the form of a slot in the starter housing through which the spring passes. These devices are assembled by winding the spring home with the sheave. Thread the movable end of the spring through the housing slot, engage the movable end with the sheave, and rotate the sheave against the direction of engine rotation until the whole length of the spring snakes through the housing slot. A notch on the end of the spring anchors it to the housing.

Rewind spring preload is necessary to maintain some rope tension when the rope is retracted. Too little preload and the rope handle droops; too much and the spring binds solid.

There are two ways to establish preload. Most manufacturers suggest the following general procedure:

1. Remove the rope handle if it is still attached.
2. Secure one end of the rope to its anchor on the sheave.
3. Wind the rope completely over the sheave, so that the sheave will rotate in the direction of engine rotation when the rope is pulled.
4. Wind the sheave against engine rotation a specified number of turns. If the specification is unknown, wind until the spring coil binds, then release the sheave for one or two revolutions.

5. Without allowing the sheave to unwind further, thread the rope through the guide tube (also called a ferrule, bushing, or eyelet) in the starter housing and attach the handle.
6. Gently pull the starter through to make certain the rope extends to its full length before the onset of coil bind and that the rope retracts smartly.

Another technique can be used when the rope anchors to the inboard (engine) side of the sheave:

1. Assemble sheave and spring.
2. Rotate the sheave, winding the mainspring until coil bind occurs.
3. Release spring tension by one to no more than two sheave revolutions.
4. Block the sheave to hold spring tension. Some designs have provisions for a nail that is inserted to lock the sheave to the housing; others can be snubbed with Vise-Grips or C-clamps.
5. With rope handle attached, thread rope through housing ferrule and anchor it to the sheave.
6. Release the sheave block and, using your thumbs for a brake, allow the sheave to rewind, pulling rope after it.
7. Test starter operation.

The starter rope should be the same weave, diameter, and length as the original. If the required length is unknown, fix the rope to the sheave, wind the sheave until coil bind—an operation that also winds the rope on sheave—and then allow the sheave to back off for one or two turns. Cut the rope, leaving enough surplus for handle attachment.

Three types of clutch assemblies are encountered: Briggs & Stratton sprag, or ratchet; Fairbanks-Morse friction-type; and the positive-engagement dog-type used by other manufacturers. In the event of slippage, clean the Briggs clutch and replace the brake springs on the other types. Fairbanks-Morse clutch dogs respond to sharpening.

One last general observation concerns starter positioning: Whenever a rewind starter has been removed from the engine or has vibrated loose, starter clutch/flywheel hub alignment must be reestablished. Follow this procedure:

1. Attach the starter or starter/blower housing assembly loosely to the engine.
2. Pull the starter handle out about 8 in. to engage the clutch.
3. Without releasing the handle, tighten the starter hold-down screws.
4. Cycle the starter a few times to check for possible clutch drag or rope bind. Reposition as necessary.

Briggs & Stratton

Briggs & Stratton side-pull starters are special in several respects (Fig. 5-4). In addition to its basic function of transmitting torque from the starter sheave to the flywheel and disengaging when the engine catches, the starter clutch also serves as the flywheel nut and starter sheave shaft. Starter and blower housing assembles are integral. It is possible, however, to drill out the spot welds and replace the starter assembly as a separate unit. Bend-over tabs locate the starter sheave in the starter housing.

Disassembly

Follow this procedure:

1. Remove blower housing and starter from engine.
2. Remove rope by cutting the knot at the starter sheave (visible from underside of blower housing).
3. Using pliers, grasp the protruding end of the mainspring and pull it out as far as possible (Fig. 5-5). Disengage the spring from the sheave by rotating the spring a quarter turn or by prying one of the tangs up and twisting the sheave.
4. Clean and inspect. Replace the rope if it is oil-soaked or frayed. Although it might appear possible to reform the end of a broken Briggs & Stratton mainspring, such efforts are in vain and the spring must be

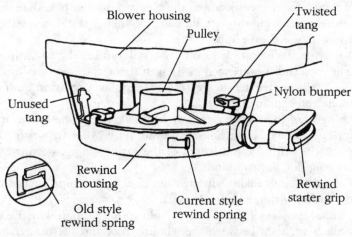

FIGURE 5-4. *Briggs & Stratton rewind starter used widely in the past and carried over today in the "Classic" line.*

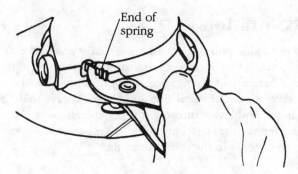

FIGURE 5-5. *Once the rope is removed, pull the rewind spring out of the starter housing. The spring can be detached from the sheave by twisting the sheave a quarter turn.*

replaced for a permanent repair. The same holds for the spring anchor slot in the housing. Once an anchor has swallowed a spring, the housing should be renewed.

Assembly

1. Dab a spot of grease on the underside of the steel sheave. Note that a plastic version requires no lubrication (Fig. 5-6).
2. Secure the blower housing engine-side up to the workbench with nails or C-clamps.
3. Working from the outside of the blower housing, pass the inner end of the mainspring through the housing anchor slot. Engage the inner end with the sheave hub.
4. Some mechanics attach rope (less handle) to the sheave at this point. The rope end is cauterized in an open flame and is knotted.
5. Bend tabs to give the sheave 1/16-in. endplay. Use nylon bushings on models so equipped.
6. Using a 1/4-in. wrench extension bar or a piece of one-by-one inserted into the sheave center hole, wind the sheave 16 turns or so counterclockwise until the full length of the mainspring passes through the housing slot and coil binds.
7. Release enough mainspring tension to align the rope anchor hole in the sheave with the housing eyelet.
8. Temporarily block the sheave to hold spring tension with a Crescent wrench snubbed between the winding tool and the blower housing (Fig. 5-7A).

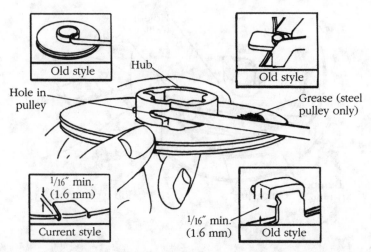

FIGURE 5-6. *Spring installation varies slightly with the date of manufacture. Steel sheaves require lubrication.*

9. If the rope has been installed, extract the end from between the sheave flanges, thread through eyelet, cauterize, and attach the handle. If the rope has not been installed, pass the cauterized end through the eyelet from outside the housing, between sheave flanges, and out through the sheave anchor hole (Fig. 5-7). Knot the end of the rope. Old-style sheaves incorporate a guide lug between flanges. The rope must pass between the lug and sheave hub. This operation is aided by a small screwdriver or a length of piano wire (Fig. 5-7A).

The clutch is not normally opened unless wear or accumulated grim causes it to slip. Older assemblies are secured with a wire retainer clip; newer versions depend upon retainer-cover tension and can be pried apart with a small screwdriver (Fig. 5-8). Clean parts with a dry rag (avoid the use of solvent). The clutch housing can be removed from the crankshaft using a special factory wrench described in Chap. 3. Assemble the unit dry, without lubricant.

Eaton

Recognizable by P-shaped engagement dogs, or pawls, Eaton starters have been used widely on American-made engines. Light-duty models employ a

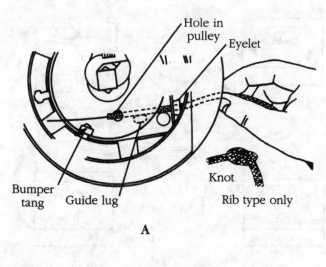

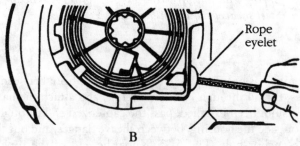

FIGURE 5-7. *Starters for most cast-iron block Briggs engines have an internal rope guide in the form of a lug buried deeply within the sheave. Use a length of piano wire to thread the rope past the inner side of the lug as shown (A). Newer designs omit the guide lug, making installation easier (B).*

single pawl (Fig. 5-9); heavier-duty models use two and sometimes three pawls. All of these starters incorporate a sheave-centering pin, usually riding on a nylon bushing.

A common complaint is failure to engage the flywheel. This difficulty can be traced to the clutch brake, which generates friction that translates into pawl engagement, or to the pawls themselves. Two brake mechanisms are encountered. The later arrangement, as shown in Figure 5-9, employs a small coil spring that reacts against the cup-like pawl retainer.

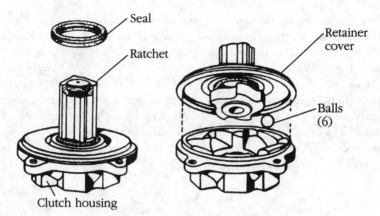

FIGURE 5-8. *Current production clutch cover is a snap fit to the clutch housing. Older versions employed a spring wire retainer. As a point of interest, older engines can be modified to accept new clutch assembly by trimming 3/8 in., from the crankshaft stub and 1/2 in. from the sheave hub.*

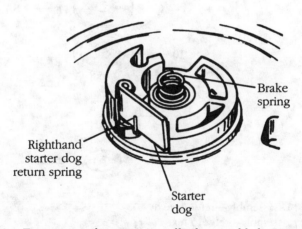

FIGURE 5-9. *Eaton rewind starter partially disassembled. Generous retainer-screw torque compresses brake spring, generating friction against the retainer that causes it to extend the dog. Because the rope attaches to the engine—and accessible—side of the sheave, the rope can be replaced by applying and holding mainspring pretension. The original rope is fished out, new rope is passed through the eyelet and sheave hole, knotted, and pretension is slowly released. As the spring uncoils, it winds rope over sheave.*

The earlier brake interposes a star-shaped washer between the pawl retainer and brake spring. Figures 5-10 and 5-11 show this part. A shouldered retainer screw secures the assembly to the sheave and preloads the brake spring (Fig. 5-12).

Check the retainer screw, which should be just short of "hernia tight"; inspect friction parts, with special attention to the optional star brake; and check the pawl return spring (Fig. 5-9), which can be damaged by engine kickbacks. Clean parts, assemble without lubricant, and observe the response of the pawls as the rope is pulled. If necessary, replace the star brake, retainer cup, and brake spring.

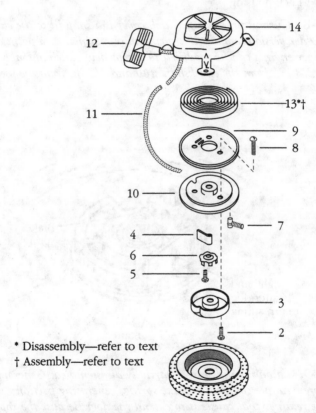

* Disassembly—refer to text
† Assembly—refer to text

FIGURE 5-10. *Eaton light-duty pattern starter used on small two- and four-stroke engines. This starter is distinguished by its uncased mainspring (13) and single-dog clutch (dog shown at 4, clutch retainer at 3). In the event of slippage during cranking, replace friction spring 5 and brake 6.*

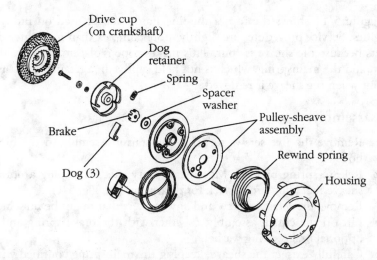

Drive cup
(on crankshaft)

Dog
retainer

Spring

Spacer
washer

Pulley-sheave
assembly

Brake

Rewind spring

Housing

Dog (3)

FIGURE 5-11. *Eaton heavy-duty starter—the type used on some Kohler engines. Note the brake friction washer, three-dog clutch, and split sheave.*

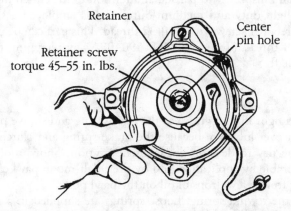

Retainer

Center
pin hole

Retainer screw
torque 45–55 in. lbs.

FIGURE 5-12. *View of the engine side of the sheave with a single-dog clutch. This unit is to be assembled dry; only snow proof models, distinguished by a half-moon cam that engages the dogs, require oil on dog-mounting posts.*

Figure 5-11 shows the top-of-the-line Eaton starter used on industrial engines. Service procedures are slightly more complex than for lighter-duty units because the sheave is split. This makes rope replacement more difficult, and the mainspring, which is not held captive in a retainer, can thrash about when the sheave is removed.

Disassembly

1. Remove the five screws securing the starter assembly to the blower housing.
2. Release spring preload. Most heavy-duty models employ a notched sheave that allows rope slack for disarming (see Fig. 5-1).
3. Remove the retainer screw and any washers that might be present.
4. Lift off the clutch assembly, together with the brake spring and the optional brake spring washer.
5. Carefully extract the sheave, keeping the mainspring confined within the starter housing.
   Warning: Wear safety glasses during this and subsequent operations.
6. Remove the rope, which may be knotted on the inboard side of the sheave or sandwiched between sheave halves as shown in Figure 5-11. The screws that hold the sheave halves together can require a hammer impact tool to loosen.
7. Remove the spring if it is to be replaced. Springs without a retainer are unwound one coil at a time from the center outward.
8. Clean and inspect with particular attention to the clutch mechanism. Older light-duty and medium-duty models employed a shouldered clutch retainer screw with a 10-32 thread. This part can be updated to a 12-28 thread (Tecumseh PN. 590409A) by retapping the sheave pivot shaft.

Assembly

1. Apply a light film of grease to the mainspring and sheave pivot shaft. Do not over lubricate because the brake spring and clutch assembly must be dry to develop engagement friction. Snowproof clutches, recognizable by application and by their half-moon pawl cam, might benefit from a few drops of oil on the pawl posts.
2. Install the rewind spring. Loose springs are supplied in a disposable retainer clip. Position the spring—observing correct engine rotation as shown in Figure 5-3—over the housing anchor pin. Gently cut the tape holding the spring to the retainer, retrieving the tape in segments. Install spring and retainer sets by simply dropping them in place.

3. Install the rope, an operation that varies with sheave construction:
   Split sheave
       A. Double-knot the rope, cauterize, and install between sheave halves, trapping the rope in the cavity provided.
       B. Install the sheave on the sheave pivot shaft, engaging the inner end of the mainspring. A punch or piece of wire can be used to snag the spring end as shown in Figure 5-13. Install the clutch assembly.
       C. Wind the sheave until the mainspring coil binds (Fig. 5-14).
       D. Carefully release spring tension two revolutions and align the rope end with the eyelet in the starter housing.
       E. Using Vise-Grips, clamp the sheave to hold spring tension and guide the rope through the eyelet. Attach the handle.
       F. Verify that sufficient pretension is present to retract rope.
   One-piece sheave
       A. Wind the sheave to coil bind and back off to align the rope hole on the inboard face of the sheave with the housing eyelet.
       B. Clamp the sheave.
       C. Cauterize the ends of the rope and install the rope through the eyelet and sheave (Fig. 5-15).

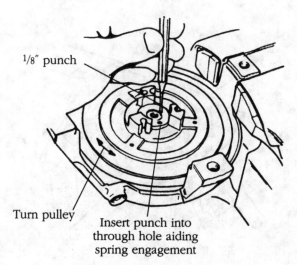

¹⁄₈″ punch

Turn pulley   Insert punch into through hole aiding spring engagement

FIGURE 5-13. *A punch aids sprint-to-sheave engagement on large Eaton starters.*

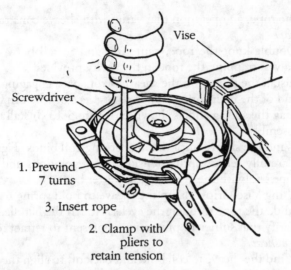

Vise

Screwdriver

1. Prewind
7 turns

3. Insert rope

2. Clamp with
pliers to
retain tension

FIGURE 5-14. *Prewind specification varies with starter model and mainspring condition.*

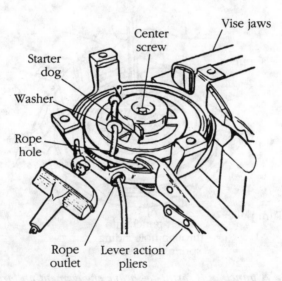

Center
screw

Vise jaws

Starter
dog

Washer

Rope
hole

Rope
outlet

Lever action
pliers

FIGURE 5-15. *Installing the rope on a one-piece sheave involves holding pretension with Vise-Grips and inserting the rope from outside the starter housing, through the eyelet, and into its anchor.*

D. Knot the rope under the sheave and install the handle.

E. Carefully release the sheave, allowing the rope to wind as the spring relaxes.

F. Test for proper pretension.

4. Pull out the centering pin (where fitted) so that it protrudes about 1/8 in. past the end of the clutch retainer screw. Some models employ a centering-pin bushing.

5. Install the starter assembly on the engine, pulling the starter through several revolutions before the hold-down screws are snubbed. Test operation.

Fairbanks-Morse

Fitted to several American engines, Fairbanks-Morse starters can be recognized by the absence of serrations on the flywheel cup. The cup is a soft aluminum casting, and friction shoes (that other manufacturers call "dogs") are sharpened for purchase. Vintage models used a wireline in lieu of the rope. Figure 5-16 is a composite drawing of Models 425 and 475, intended for large single-cylinder engines.

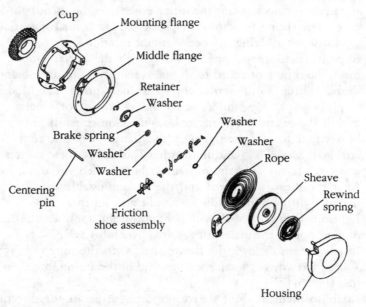

FIGURE 5-16. *Fairbanks-Morse starter used on Kohler and other heavy-duty engines. Mounting and middle flanges are characteristic of Model 475.*

Disassembly

1. Remove the starter assembly from the blower housing.
2. Turn the starter over on bench and, holding the large washer down with thumb pressure, remove the retainer ring that secures the sheave and clutch assembly (Fig. 5-17A).
3. Remove the washer, brake spring, and friction shoe assembly. Normally, the friction shoe assembly is not broken down further.
4. Relieve mainspring preload by removing the rope handle and allowing the sheave to unwind in a controlled fashion. Tension on the Model 475 can be released by removing the screws holding the middle and mounting flanges together (Fig. 5-17B).
5. Cautiously lift the sheave about 1/2 in. out of the housing and detach the inner spring end from the sheave hub.
6. Leave the mainspring undisturbed (unless you are replacing it). To remove the spring, lift one coil at a time, working from the center outward. Wear eye protection.
7. Clean all parts in solvent and inspect.

Assembly

1. Install the spring, hooking the spring eyelet over the anchor pin on the cover. The spring lay shown in Figure 5-17D is for conventional—clockwise when facing flywheel—engine rotation.
2. Rope installation and preload varies with the starter model. In all cases, the rope is attached to the sheave and wound on it before the sheave is fitted to the starter cover and mainspring. The Model 475 employs a split rope guide, or ferrule, consisting of a notch in the middle flange and in the starter housing. Consequently, the rope can be secured to and wound over the flange with the rope handle attached. Model 425 and most other Fairbanks-Morse starters use a one-piece ferrule and the rope must be installed without a handle. After the sheave is secured and the preload established, the rope is threaded through the ferrule for handle attachment.
3. Lubricate the sheave shaft with light grease and apply a small quantity of motor oil to the mainspring. Avoid over lubrication.
4. Install the sheave over the sheave shaft with the rope fully wound. With a screwdriver, hook the inner end of the spring into the sheave hub (Fig. 5-17E).
5. Establish preload—four sheave revolutions against the direction of engine rotation for Model 425, five turns for Model 475, and variable for others.

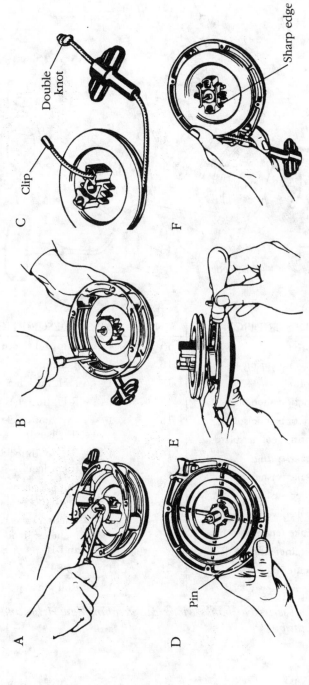

FIGURE 5-17. *Crucial service operations include removing the retainer ring and spring-loaded washer (A), releasing residual spring tension (B), rope anchors and rope lay for standard engine rotation (C), mainspring orientation for standard rotation (D), spring and sheave engagement (E), and correct brake-shoe assembly (F).*

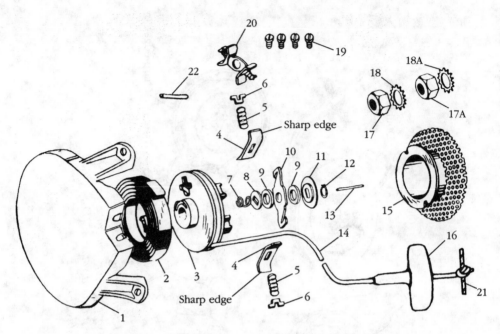

ILLUS. NO.	QTY.	DESCRIPTION	ILLUS. NO.	QTY.	DESCRIPTION
1	1	Cover	15	1	Cup and screen
2	1	Rewind spring	16	1	T-handle
3	1	Rotor	17	1	L.H. thick hex nut
4	2	Friction shoe plate	17A	1	R.H. thick hex nut
5	2	Friction shoe spring	18	1	Ext. tooth lockwasher (left hand)
6	2	Spring retainer plate	18A	1	Ext. tooth lockwasher (right hand)
7	1	Brake spring	19	4	Pan hd. Screw w/int.-ext. tooth lockwasher
8	1	Brake washer			
9	2	Fiber washer	20	1	Friction shoe assembly, includes: Items 4, 5, 6, and 10
10	1	Brake lever			
11	1	Brake retainer washer			
12	1	Retainer ring	21	1	Spiral pin
13	1	Centering pin	22	1	Roll pin
14	1	Cord			

FIGURE 5-18. *Small series Fairbanks-Morse can accommodate right- and left-hand engine rotation.*

6. Complete the assembly, installing sheave hold-down hardware and the friction-shoe assembly. When assembled correctly, the sharp edges of the friction shoes are poised for contact with the flywheel-hub inside diameter (Fig. 5-17F).
7. Pull the centering pin out about 1/8 in. for positive engagement with the crankshaft center hole.
8. Install the starter on the blower housing, rotating the flywheel with the starter rope as the hold-down screws are torqued. This procedure helps to center the clutch in the flywheel hub.
9. Start the engine to verify starter operation.

The Fairbanks-Morse utility starter is a smaller and simpler version of the heavy-duty models just discussed (Fig. 5-18). A one-piece sheave is used with the rope anchored by a knot, rather than a compression fitting. The utility starter uses the same clutch components as its larger counterparts and, like them, can be assembled for right- or left-hand engine rotation.

Vertical pull

Vertical-pull starters are an area where DIY mechanics shine. These starters are obsolete, complicated, and troublesome. Commercial shops usually won't fool with them and when they do, the labor charges can be horrendous. But a do-it-yourselfer can, with a bit of patience, repair these starters and, in the process, salvage engines that would otherwise be scrapped.

Like other spring-powered devices, these starters must be disarmed before disassembly. Otherwise, the starter will disarm itself with unpredictable results. Disarming involves three distinct steps: releasing mainspring pretension (usually by slipping a foot or so of rope out of the sheave flange and allowing the sheave to unwind), disengaging the mainspring anchor (usually held by a threaded fastener), and when the spring is to be replaced, uncoiling the spring from its housing.

Warning: Safety glasses are mandatory for disassembly.

Vertical-pull starters tend to be mechanically complex and—because of a heavy reliance upon plastic, light-gauge steel, and spring wire—are unforgiving. Parts easily bend or break. Lay components out on the bench in proper orientation and in sequence of disassembly. If there is any likelihood of confusion, make sketches to guide assembly. Also, note that the step-by-step instructions in this book must aim at thoroughness and describe all operations, but it will rarely be necessary to follow every step and completely dismantle a starter.

Briggs & Stratton

Briggs & Stratton has used one basic vertical-pull starter with minor variations in the link and sheave mechanisms. It is probably the most reliable of
these starters, and the easiest to repair.

Disassembly

1. Remove starter assembly from the engine.
2. Release mainspring pretension by lifting the rope out of the sheave
    flange and, using the rope for purchase, winding the sheave counter-
    clockwise two or three revolutions (Fig. 5-19).
3. Carefully pry the plastic cover off with a screwdriver. Do not pull on
    the rope with the cover off and spring anchor attached; doing so can
    permit the outer end of the spring to escape the housing.
4. Remove the spring anchor bolt and spring anchor (Fig. 5-20). If the
    mainspring is to be replaced, carefully extract it from the housing,

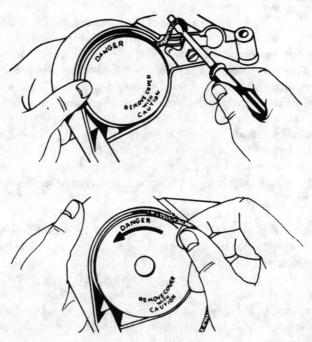

FIGURE 5-19. *Briggs & Stratton vertical-pull starters are disarmed by slipping
the rope out of the sheave groove and using the rope to turn the sheave two or
three revolutions counterclockwise until the mainspring relaxes.*

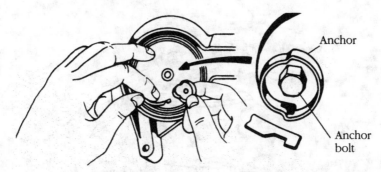

FIGURE 5-20. *Mainspring anchor bolt must be torqued 75 lb/in. and can be further secured with thread adhesive.*

working from the center coil outward. Note the spring lay for future reference.

5. Separate the sheave and the pin (Fig. 5-21). Observe the link orientation.

6. The rope can be detached from the sheave with the aid of long-nosed pliers. Figure 5-22 shows this operation and link retainer variations.

7. The rope can be disengaged from the handle by prying the handle center section free and cutting the knot (Fig. 5-23).

8. Clean all parts (except rope) in petroleum-based solvent to remove all traces of lubricant.

9. Verify the gear response to link movement as shown in Figure 5-24. The gear should move easily between its travel limits. Replace the link as necessary.

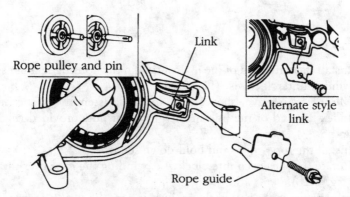

FIGURE 5-21. *Make note of the friction-link orientation for assembly.*

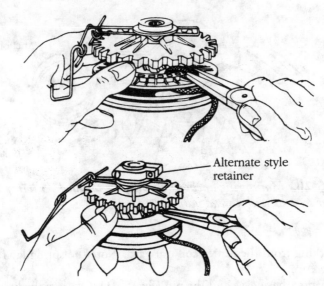

FIGURE 5-22. *The rope can be disengaged from the sheave with long-nosed pliers.*

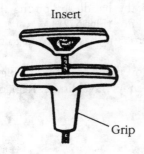

FIGURE 5-23. *The Briggs handle employs an insert that must be extracted to renew the rope.*

Assembly

1. Install the outer end of the mainspring in the housing retainer slot and wind counterclockwise (Fig. 5-25).
2. Mount the sheave, sheave pin, and link assembly in the housing. Index the end of the link in the groove or hole provided (Figs. 5-26 and 5-27).
3. Install the rope guide and hold-down screw.
4. Rotate the sheave counterclockwise, winding the rope over the sheave (Fig. 5-28).
5. Engage the inner end of the mainspring on the spring anchor. Mount the anchor and torque the hold-down caps crew to 75–90 lb/in.

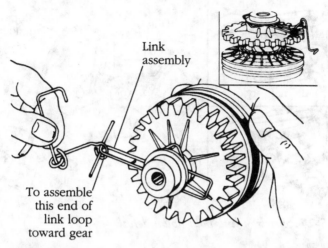

Link
assembly

To assemble
this end of
link loop
toward gear

FIGURE 5-24. *The pinion gear should move through its full range of travel in response to link movement. The inset on the upper right of the illustration shows link orientation.*

Spring
retainer
slot

FIGURE 5-25. *The mainspring winds counterclockwise from the outer coil.*

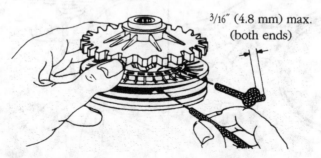

3/16″ (4.8 mm) max.
(both ends)

FIGURE 5-26. *A short length of piano wire aids rope insertion into the sheave.*

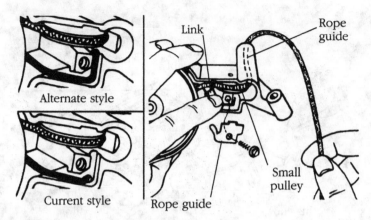

FIGURE 5-27. *Friction link hold-down detail.*

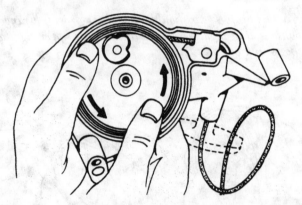

FIGURE 5-28. *The rope winds counterclockwise on the sheave, then the spring anchor and anchor bolt are installed.*

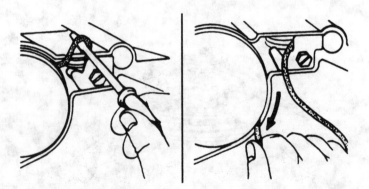

FIGURE 5-29. *Pretension requires two or three sheave revolutions using the rope for leverage.*

6. Snap the plastic cover into place over the spring cavity.
7. Disengage 12 in. or so of rope from the sheave and, using the rope for purchase, turn the sheave two or three revolutions clockwise to generate pretension (Fig. 5-29).
8. Mount the starter on the engine and test.

Tecumseh

Tecumseh has used several vertical-pull starters, ranging from quickie adaptations of side-pull designs to the more recent vertical-engagement type.

The gear-driven starter shown in Figure 5-30 is an interesting transition from side to vertical-pull. No special service instructions seem appropriate, except to provide plenty of grease in the gear housing and some light lubrication on the mainspring. Assemble the brake spring without lubricant.

The current horizontal engagement starter (Fig. 5-31) is reminiscent of the Briggs & Stratton design, with a rope clip, cup-type spring anchor ("hub" in the drawing), and threaded sheave extension upon which the pinion rides.

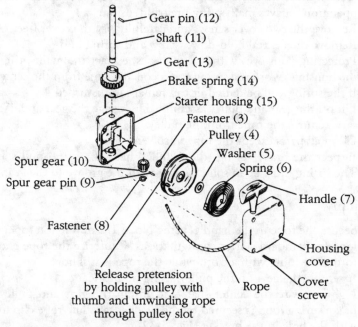

FIGURE 5-30. *Early Tecumseh vertical-pull starter, driving through a gear train. While heavy (and, no doubt, expensive to manufacture), this starter was quite reliable.*

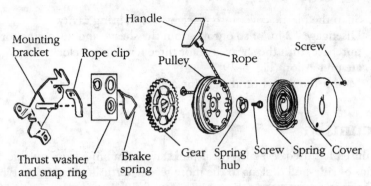

Handle

Mounting
bracket   Rope clip   Screw
                    Pulley   Rope

Thrust washer   Brake   Gear Spring Screw Spring Cover
and snap ring   spring   hub

FIGURE 5-31. *Tecumseh's most widely used vertical-pull starter employed a spiral gear to translate the pinion horizontally into contact with the flywheel.*

Disassembly

1. Remove the unit from the engine.
2. Detach the handle and allow the rope to retract past the rope clip. This operation relieves mainspring preload tension.
3. Remove the two cover screws and carefully pry the cover free.
4. Remove the central hold-down screw and spring hub.
5. Protecting your eyes with safety glasses, extract the mainspring from the housing. Work the spring free a coil at a time from the center out. If the spring will be reused, it can remain undisturbed.
6. Lift off the gear and pulley assembly. Disengage the gear and, if necessary, remove the rope from the pulley.
7. Clean all parts except the rope in solvent.
8. Inspect the brake spring (the Achilles' heel of vertical-pull starters). The spring must be in solid contact with the groove in the gear.

Assembly

1. Secure the rope to the handle, using No. 4 1/2 or 5 nylon rope, 61 in. long for standard starter configurations. Cauterize the rope ends and form by wiping with a cloth while the rope is still hot.
2. Assemble the gear on the pulley, using no lubricant.
3. Lightly grease the center shaft and install the gear and pulley. The brake spring loop is secured by the bracket tab. The rope clip indexes with the hole in the bracket (Fig. 5-32).

4. Install the hub and torque the center screw to 44–55 lb/in.
5. Install the spring. New springs are packed in a retainer clip to make installation easier.
6. Install the cover and cover screws.
7. Wind the rope on the pulley by slipping it past the rope clip. When fully wound, turn the pulley two additional revolutions for preload.
8. Mount the starter on the engine, adjusting the bracket for minimum 1/16-in. tooth clearance (Fig. 5-33). Less clearance could prevent disengagement, destroying the starter.

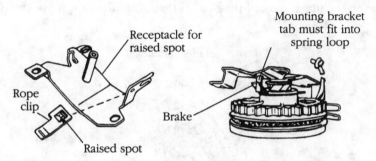

FIGURE 5-32. *The rope clip and spring loop index to the bracket.*

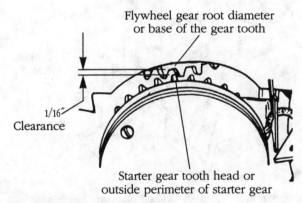

FIGURE 5-33. *Generous gear lash, minimum 1/16 in., is required to assure pinion disengagement when the engine starts.*

Vertical pull, vertical engagement

The vertical-pull, vertical-engagement starter is a serious piece of work that demands special service procedures. It is relatively easy to disassemble while still armed. The results of this error can be painful. Another point to note is that rope-to-sheave assembly as done in the field varies from the original factory assembly.

Rope replacement

Figure 5-34 is a composite drawing of several vertical-pull starters. Many do not contain the asterisked parts, and early models do not have the V-shaped groove on the upper edge of the bracket that simplifies rope replacement.

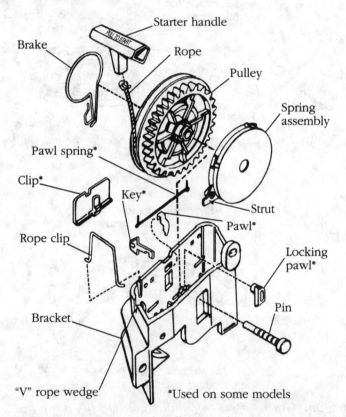

FIGURE 5-34. *Tecumeh's vertical-pull, vertical-engagement starter is the most sophisticated unit used on small engines. The mainspring and its retainer are integral and are not separated for service.*

When this groove is present, the rope (No. 4 1/2, 65-in. standard length, longer with a remote rope handle) can be renewed by turning the sheave until the staple, which holds the rope to the sheave, is visible at the groove (Fig. 5-35). Pry out the staple and wind the sheave tight. Release the sheave a half turn to index the hole in the sheave with the V-groove. Insert one end of the replacement rope through the hole, out through the bracket. Cauterize and knot the short end, and pull the rope through, burying the knot in the sheave cavity. Install the rope handle, replacing the original staple with a knot, and release the sheave. The rope should wind itself into place.

Disassembly

1. Remove the starter from the engine.
2. Pull out the rope far enough to secure it in the V-wedge on the bracket end. This part, distinguished from the V-groove mentioned above, is called out in Figure 5-34.
3. The rope handle can be removed by prying out the staple with a small screwdriver.
4. Press out the flat-headed pin that supports the sheave and spring the capsule in the bracket. This operation can be done in a vise with a large deep well socket wrench as backup.
5. Turn the spring capsule to align with the brake spring legs. Insert a nail or short (3/4-in. long maximum) pin through the hole in the strut and into the gear teeth (Fig. 5-36).

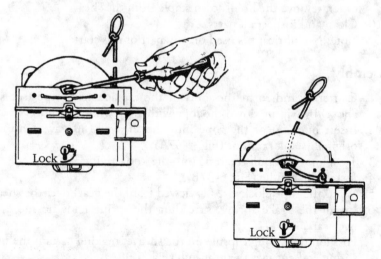

FIGURE 5-35. *V-groove in the bracket gives access to the rope anchor on some models.*

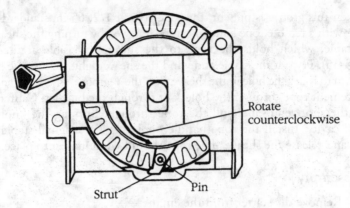

FIGURE 5-36. *A pin locks the spring capsule and gear to prevent sudden release of the mainspring tension.*

6. Lift the sheave assembly and pry the capsule out of the bracket. Warning: Do not separate the sheave assembly and capsule until the mainspring is completely disarmed.
7. Hold the spring capsule firmly against the outer edge of the sheave with thumb pressure and extract the locking pin inserted in Step 5.
8. Relax pressure on the spring capsule, allowing the capsule to rotate, thus dissipating residual mainspring tension.
9. Separate the capsule from the sheave and, if rope replacement is in order, remove the hold-down staple from the sheave.
10. Clean and inspect all parts.
    Note: No lubricant is used on any part of this starter.

Assembly

1. Cauterize and form the ends of the replacement rope (see specs above) by wiping down the rope with a rag while still hot.
2. Insert one end of the rope into the sheave, 180° away from the original (staple) mount (Fig. 5-37A).
3. Tie a knot and pull the rope into the knot cavity.
4. Install the handle (Fig. 5-37B).
5. Wind the rope clockwise (as viewed from the gear) over the sheave.
6. Install the brake spring, spreading the spring ends no more than necessary.
7. Position the spring capsule on the sheave, making certain the mainspring end engages the gear hub (Fig. 5-38A).
8. Wind four revolutions, align the brake spring ends with the strut (Fig. 5-38B), and lock with the pin used during disassembly.

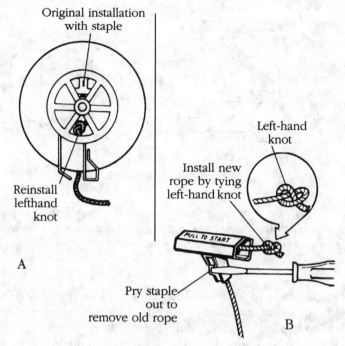

FIGURE 5-37. *Replacement ropes anchor with a knot, rather than staple, and mount 180° from the original position on the sheave.*

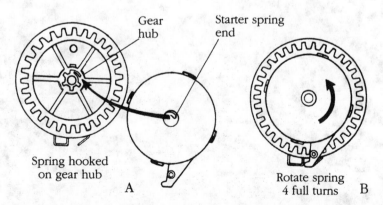

FIGURE 5-38. *The spring capsule engages with the gear hub (A), is rotated four revolutions for pretension, and pinned (B).*

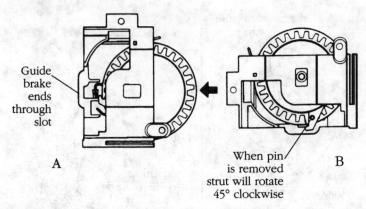

Guide brake ends through slot

A

When pin is removed strut will rotate 45° clockwise

B

FIGURE 5-39. *The sheave and spring capsule assembly installs in the bracket with the brake spring ends in slots (A). Releasing pin arms starter (B), which can now be mounted on the engine.*

9. Install pawls, springs, and other hardware that might be present.
10. Insert the sheave and spring assembly into the bracket, with the brake spring legs in the bracket slots (Fig. 5-39).
11. Feed the rope under the guide and snub it in the V-notch.
12. Remove the locking pin, allowing the strut to rotate clockwise until retained by bracket.
13. Press or drive the center pin home.
14. Mount the starter on the engine and test.

6

Electrical system

At its most developed, the electrical system consists of a charging circuit, a storage battery, and a starting circuit. A flywheel alternator provides electrical energy that is collected in the battery for eventual consumption by the starter motor.

Not all small engine electrical systems include both circuits. Some dispense with the starting circuit and others employ a starting circuit without provision for onboard power generation.

Starting circuits

Starting circuits fall into two major groups: dc (direct current) systems that receive power from a 6 or 12 V battery and ac (alternating current) systems that feed from an external 120 VAC line. I will not discuss ac systems because the hazards implicit in line-current devices cannot be adequately addressed in a book of this type. My discussion is limited to dc systems that employ conventional (lead-acid) or NiCad batteries.

Lead acid

As shown in Figures 6-1 and 6-2, a conventional starting circuit includes four major components—battery, ignition switch, solenoid, and motor—wired into two circuit loops:

- Control loop—14-gauge primary wire from the positive battery terminal, through the ignition switch, to the solenoid windings.
- Power loop—cable from the positive battery terminal, through the solenoid and to the starter motor.

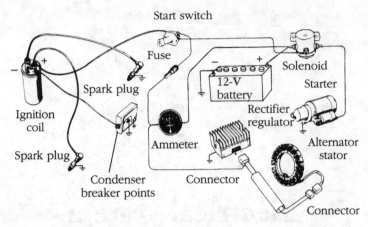

Start switch

Fuse

Spark plug

Ignition coil

Spark plug

Condenser breaker points

12-V battery

Solenoid

Starter

Rectifier regulator

Ammeter

Connector

Alternator stator

Connector

FIGURE 6-1. *Electrical system supplied with Onan engines and tied into battery-and-coil ignition. Note the heavy-duty stator and combined rectifier/regulator.*

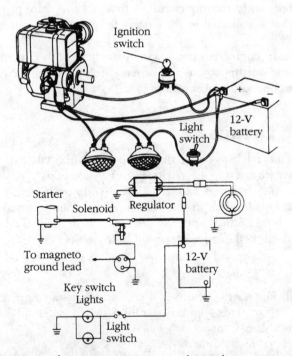

Ignition switch

Light switch

12-V battery

Starter

Solenoid

Regulator

To magneto ground lead

12-V battery

Key switch Lights

Light switch

FIGURE 6-2. *Briggs & Stratton 10A system shows the tie-in to a charging system on the positive battery post. The ignition switch grounds magneto and does not interchange with automotive-type switches.*

As shown in the drawings, the negative battery terminal connects to the engine block to provide a ground return for both loops. When energized, current flows from the positive side of the solenoid, starter motor, and other circuit components to ground.

The solenoid—more properly called a relay—is a normally open (NO) electromagnetic switch. When energized by the starter switch, the solenoid closes with an audible click to complete the power circuit. Most solenoids are internally grounded, which means that mounting faces must be clean and hold-down bolts secure. Interlocks are sometimes included in the control circuit to prevent starting under unsafe conditions (e.g., if the machine is in gear). The charging system delivers current to the positive post of the battery.

Figure 6-3 outlines diagnostic procedures which use the solenoid as the point of entry. Shunting the solenoid with a jumper cable removes the solenoid and the control loop from the circuit to give an immediate indication of starter motor function.

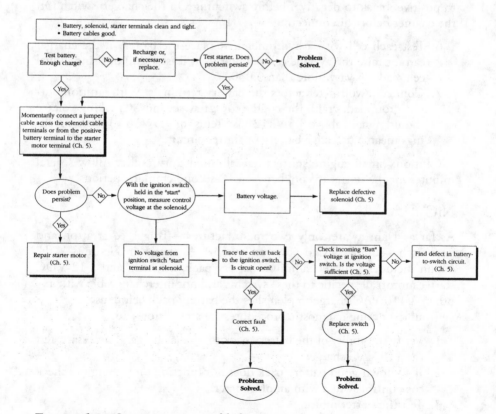

FIGURE 6-3. *Starter circuit troubleshooting.*

The majority of starting-circuit malfunctions are the fault of the battery. Loose, dirty, or corroded connections account for most of the others.

1. Remove the cable connections at the battery and scrape the battery terminals to bright metal. Repeat the process for solenoid and starter-motor connections.
2. Clean the battery connection at the engine and verify that all circuit components are bolted down securely to the engine or equipment frame.
3. Verify that electrolyte covers the battery plates. Add distilled water as necessary, but do not overfill.
4. Clean the battery top with a mixture of baking soda, detergent, and water. Rinse with fresh water and wipe dry.
5. With the battery cables disconnected, charge the battery.

Warning: Lead-acid batteries give off hydrogen gas during charging. To minimize sparking and possible explosion, connect the charger cables (red to positive, black to negative) before switching On the charger; switch Off the charger before disconnecting.

6. Test each cell with a hydrometer. Replace the battery if the charger cannot raise the average cell reading to at least 1.260 or if individual cell readings vary more than 0.050.
7. Connect a voltmeter across the battery terminals. With ignition output grounded, crank the engine for a few seconds. Cranking voltage should remain above 9.5 V (12 V systems) or 4.5 V (6 V). Lower readings mean a defective battery or starter circuit.

Caution: Small-engine starters have limited duty cycles. Allow several minutes for the starter to cool between 10-second cranking periods.

NiCad

As far as I am aware, only two manufacturers—Briggs & Stratton and Tecumseh—supply NiCad-powered systems. As shown in Figure 6-4, the circuit includes a nickel-cadmium battery pack, a switch, and a 12 VDC starter motor, all specific to these systems and not interchangeable with any other. A 110 VAC charger replenishes the battery pack before use.

Troubleshooting diagnostic procedures are straightforward:

1. Verify the ability of the battery pack to hold a charge. If necessary, test the 110 VAC charger.
2. If a known-good battery pack does not function on the engine, check the control switch with an ohmmeter.
3. Test the starter motor.

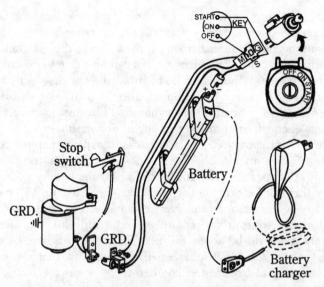

FIGURE 6-4. *Briggs & Stratton Nicad system integrates engine controls with the wiring harness. Tecumseh's practice is similar.*

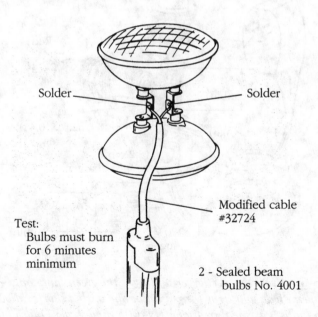

FIGURE 6-5. *Nicad test load is fabricated from two sealed-beam headlamps and a battery-to-starter cable.*

Refusal of the battery pack to hold a charge is the most common fault. After 16 hours on the charger, battery potential should range between 15.5 V and 18 V. Assuming that voltages fall within these limits, the next step is to test capacity through a controlled discharge. If a carbon-pile tester is not available, connect two No. 4001 headlamp bulbs in parallel, as shown in Figure 6-5. A freshly charged battery pack should illuminate the lamps brightly for five minutes (Briggs & Stratton) or six minutes (Tecumseh), figures that represent enough energy to start the average engine about 30 times.

Warning: Dispose of NiCad battery packs in a manner approved by local authorities. Cadmium, visible as a white powder on leaking cells, is a persistent poison. Do not incinerate and/or weld near the battery pack.

Battery life will be extended if charging is limited to 12 or 16 hours immediately before use and once every two months in dead storage.

NiCad charger output varies with battery condition, but after two or three hours it should be about 80 mA. Tecumseh lists a test meter (PN 670235) for the PN 32659 charger; Briggs suggests that the technician construct a tester, as described in Figure 6-6.

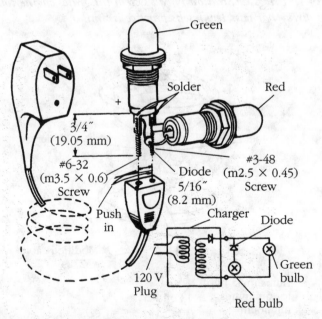

FIGURE 6-6. *A functional charger will light the green lamp only. A charger with an open diode will light the red bulb; one with a shorted diode will light both bulbs. Parts required: one 1N4005 diode, two Dialco lamp sockets (PN 0931-102 red, PN 0932-102 green), two No. 53 bulbs, and hold-down screws.*

Starter motors

Industrial motors, such as the Prestolite unit shown in Figure 6-7, are rebuildable and can be serviced by most automotive electrical shops. The Briggs & Stratton motor, shown in Figure 6-8, is also rebuildable (thanks to adequate parts support) and can, in the context of small engines, be considered a heavy-duty motor. American Bosch, used by Kohler and shown in Figure 6-9, is, from a reparability point of view, on a par with the Briggs. European Bosch, Bendix, Nippon Denso, and Mitsubishi starter motors are of similar quality. Light-duty units, such as the NiCad starter shown in Figure 6-10, do not justify serious repair efforts.

Troubleshooting

Figure 6-11 describes motor failure modes and likely causes. References to field failures apply only to those motors that use wound field coils; PM (permanent magnet) fields are, of course, immune to electrical malfunction.

Repairs

With the exception of replacing the inertial clutch, repair procedures discussed here apply to heavy-duty motors. Upon disassembly, clean the interior of the starter with an aerosol product intended for this purpose. Do not use a petroleum-based solvent. Note the placement of thrust and insulating washers.

- Inertial clutch—Shown clearly in Figure 6-10 and tangentially in other drawings, the Bendix is serviced as a complete assembly. It secures to the motor shaft with a nut, spring, clip, or roll pin. Support the free end of the motor shaft when driving the pin in or out. The helix and gear install dry, without lubrication; the pinion ratchet can be lightly oiled.
- End cap—Scribe the end cap and motor frame as an assembly aid. Installation can be tricky when the brush assembly is part of the cap. In some cases, you can retract the brushes with a small screwdriver. Radially deployed brushes can be retained with a fabricated bracket (Fig. 6-12).
- Bushings—Do not disturb the bushings unless replacements are at hand. Drive out the pinion-end bushing with a punch sized to the bushing outside diameter. Commutator bushings must be lifted out of their blind end-cap bosses, which can be done by filling the cavity with grease and using a punch, sized to match the shaft outside diameter as a piston. Hammer the punch into the grease.
   Sintered bronze bushings—Recognizable by their dull, sponge-like appearance, these bushings should be submerged in motor oil for a few minutes before installation. Brass bushings require a light temperature-resistant grease, such as Lubriplate.

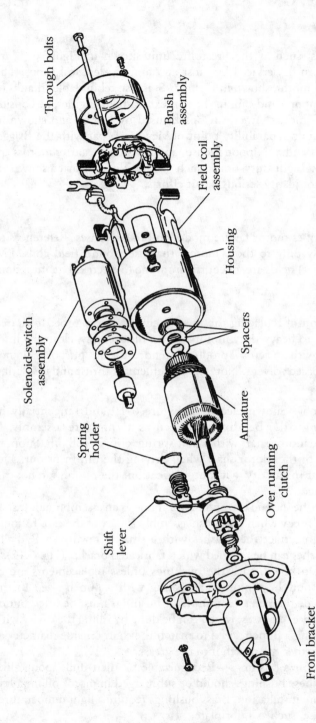

FIGURE 6-7. *Onan-supplied starter, used on some twin-cylinder applications, is the real McCoy. Rather than the conventional inertial clutch, this starter engages the pinion with a solenoid. Test the solenoid by connecting a jumper from the solenoid battery terminal to the solenoid motor terminal.*

Through bolts

Brush
assembly

Field coil
assembly

Housing

Spacers

Armature

Over running
clutch

Solenoid-switch
assembly

Spring
holder

Shift
lever

Front bracket
assembly

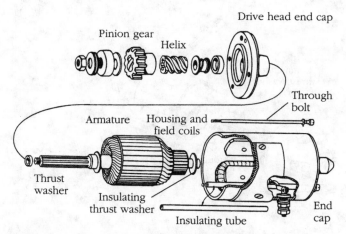

FIGURE 6-8. *Briggs & Stratton 12 VDC starter motor employs electromagnetic (EM) fields, a thrust washer on the drive end, and an insulating thrust washer at the commutator.*

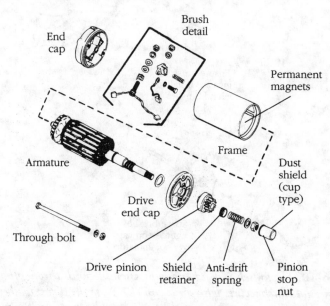

FIGURE 6-9. *American Bosch 12 VDC starter features permanent magnet (PM) fields and a radial commutator with brushes parallel to the motor shaft. Used on Kohler and other serious engines.*

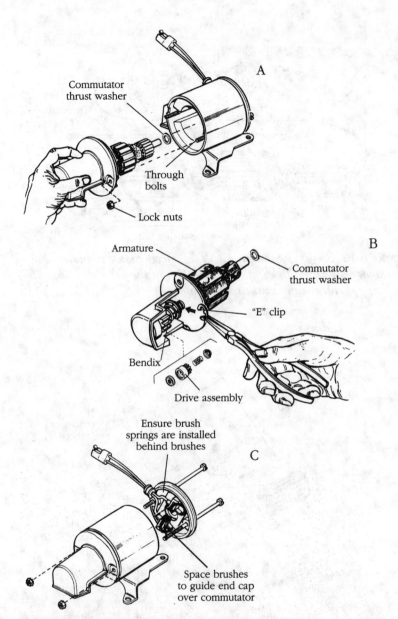

FIGURE 6-10. *Tecumseh NiCad end caps assemble with through bolts (A). Bendix inertial clutch secures with an E-clip (B). Note the thrust washer (C). Brushes, replaceable only as part of the cap assembly, must be shoehorned over the commutator. This particular starter should draw 20A while turning the engine 415 rpm with crankcase oil at room temperature.*

It is assumed that full battery voltage reaches the starter and that the flywheel offers only normal resistance to turning.

Starter does not function:

- Brushes stuck in holders
- Dirty, oily brushes/commutator
- Open, internally shorted, or grounded field coil
- Open, internally shorted, or grounded armature

Starter cranks slowly
(minimal acceptable cranking speed = 350 rpm):

- Worn brushes or weak brush springs
- Dirty, oily, or worn commutator
- Worn shaft bushings
- Defective armature

Starter stalls under compression:

- Overly advanced ignition timing
- Defective armature
- Defective field coil

Starter works intermittently:

- Sticking brushes
- Loose connections in external circuit
- Dirty, oily commutator

Starter spins freely without turning flywheel:

- Pinion gear sticking on shaft
- Broken pinion and/or flywheel teeth

FIGURE 6-11. *Starter motor faults and probable causes.*

- Brushes—Most starter problems originate with brushes that wear short and bind against the sides of their holders. As a rule of thumb, replace the brushes when worn to half their original length. Older starters used screw-type brush terminals; newer starters employ silver solder or integrate brushes and brush holders with the cap. Note the lay of the brushes: Rubbing surfaces must conform to the convexity of the commutator.
- Commutator—Heavy-duty starters can be "skimmed" on a lathe to restore commutator concentricity and surface finish. A light-duty commutator might benefit from polishing with 000-grade sandpaper. Do not use emery cloth.

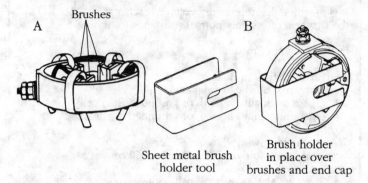

A   Brushes   B

Sheet metal brush
holder tool

Brush holder
in place over
brushes and end cap

FIGURE 6-12. *Fabricated brush holders for four-pole (brush) assemblies (A) and two-pole (B).*

- Armature—Check continuity with an ohmmeter or 12 V trouble light. Two conditions must be met:
  (1) Paired commutator bars are connected in series and should provide a continuous circuit path from a brush on one side of the commutator to its twin on the other side.
  (2) No pair of bars has continuity with other pairs or with the motor shaft (Fig. 6-13).
  Automotive electrical shops can test the armature for internal shorts with a growler.

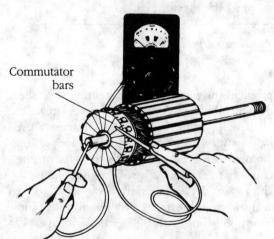

Commutator
bars

FIGURE 6-13. *Testing an American Bosch starter commutator for shorts to the motor shaft.*

- Fields—Inspect PM fields for mechanical damage (from contact with the armature) and for failure of the adhesive backing. A specialist should test electromagnetic fields, although at this point you have reached the end of practical reparability for even the best starter motor.

Charging circuits

In its most vestigial form, a charging circuit consists of a coil, flywheel magnet, and a load, such as a headlamp. Coil output alternates, or changes direction, each time a flywheel magnet excites it (the discussion of magneto theory in Chap. 3 explains why). Voltage is speed-sensitive: at idle the lamp barely glows; at wide-open throttle the filament verges on self-destruct.

Adding a battery means that stator output must be rectified, or converted from ac to dc. This is almost always done by means of one or two silicon diodes, which act as check valves to pass current flowing in one direction and block it in the other. Single-diode rectifiers pass that half of stator output that flows in the favored direction (Fig. 6-14A). Full-wave rectifiers use two diodes, wired in a bridge circuit, to impose unidirectionality upon all of the output, so that none of it goes to waste (Fig. 6-14B).

The battery receives a charge so long as its terminal voltage is lower than rectifier output voltage. The battery also acts as a ballast resistor, limiting output voltage and current. Even so, these values remain closely tied to engine speed and not electrical loads.

More sophisticated circuits use a solid-state regulator to synchronize charging current and voltage with battery requirements. The regulator caps voltage output at about 14.7 V and responds to low battery terminal voltage with more current.

The usual practice is to encapsulate the regulator with the rectifier. Look for a potted "black box" or a finned aluminum can, mounted under or on the engine shroud. All of these units share engine ground with the stator and battery. Hold-down bolts must be secure and mating surfaces clean.

Most regulator/rectifiers have three wires going to them, as shown in Figures 6-2 and 6-15. Two of these wires carry ac from the stator and one conveys B+ voltage to the battery. Wires that supply B+ are often, but not always, color-coded red.

Twenty and 30A systems can include a bucking coil (a kind of electrical brake) to limit output. The presence of such a coil is signaled by one (or sometimes two) additional wires from the stator to regulator-rectifier.

Use a high-impedance meter, preferably digital, for voltage checks. Identify circuits before testing, with particular attention to the magneto primary, which is often integrated into the regulator-rectifier connector. That circuit carries some 300 VAC.

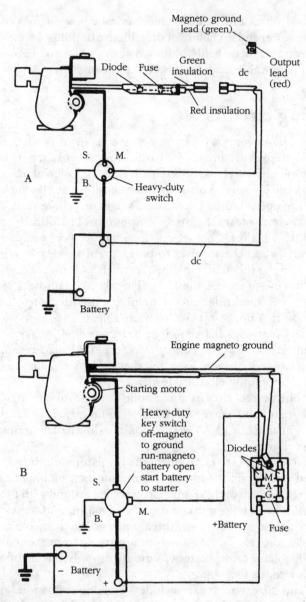

FIGURE 6-14. *Tecumseh 3A systems illustrate two approaches to rectification. The single-diode, half-wave rectifier, located in wiring harness, passes half of stator output to battery (A). The two-diode, full-wave rectifier utilizes all of stator output, doubling the charge rate (B). In the event of overcharging, one diode can be removed.*

Do not:

- Reverse polarity—reversed battery or jumper cable connections will ruin the regulator/rectifier on all but the handful of systems that incorporate a blocking diode.
- Introduce stray voltages—disconnect the B+ rectifier-to-battery lead before charging the battery or arc welding.
- Create direct shorts—do not ground any wire or touch ac output leads together.
- Operate the system without a battery—when open-circuit, unregulated ac output tests are permitted, make them quickly at the lowest possible engine rpm/voltage needed to prove the stator.
- Run the engine without the shroud in place—if necessary, route test leads outside of the shroud.

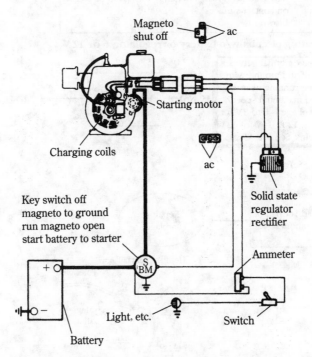

FIGURE 6-15. *Most regulator/rectifiers have three terminals—AC, AC, B+—and ground to the engine through the hold-down bolts.*

Figure 6-16 presents a standard troubleshooting format used by many small engine mechanics. It applies to all unregulated systems and to more than 90 percent of systems with a regulator or regulator-rectifier. There are exceptions: Certain regulators and regulator/rectifiers do not tolerate hot (engine running) disconnects. These components will be damaged by attempts to measure open-circuit ac voltage.

An extensive inquiry has uncovered two of these maverick systems; there are almost certainly others among the thousands of models and types of small engines sold in the United States (Kawasaki, for example, lists more than 300 distinct models).

> • General—all systems
>
> Check for: Broken or shorted wires (use ohmmeter for
> in-harness tests).
> Loose or corroded connections (pay special
> attention to pin and blade type connectors).
> Bad grounds (loose mounting bolts, paint or grease
> on contact surfaces).
> Blown fuse.

ac generating circuits-w/o battery or regulator, 6 or 12 V

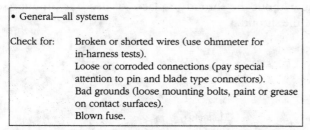

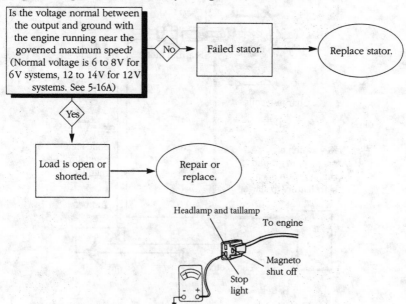

FIGURE 6-16. *Charging system troubleshooting.*

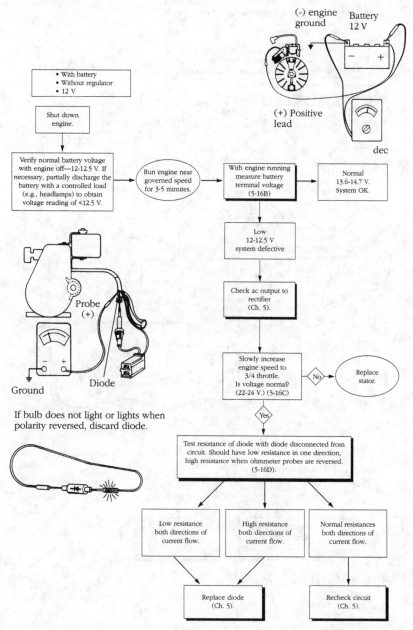

- With battery
- Without regulator
- 12 V

Shut down engine.

(-) engine ground

Battery 12 V

(+) Positive lead

dec

Verify normal battery voltage with engine off—12-12.5 V. If necessary, partially discharge the battery with a controlled load (e.g., headlamps) to obtain voltage reading of <12.5 V.

Run engine near governed speed for 3-5 minutes.

With engine running measure battery terminal voltage (5-16B)

Normal 13.6-14.7 V. System OK.

Low 12-12.5 V system defective

Check ac output to rectifier (Ch. 5).

Probe (+)

Ground

Diode

If bulb does not light or lights when polarity reversed, discard diode.

Slowly increase engine speed to 3/4 throttle. Is voltage normal? (22-24 V.) (5-16C)

No

Replace stator.

Yes

Test resistance of diode with diode disconnected from circuit. Should have low resistance in one direction, high resistance when ohmmeter probes are reversed. (5-16D).

Low resistance both directions of current flow.

High resistance both directions of current flow.

Normal resistances both directions of current flow.

Replace diode (Ch. 5).

Recheck circuit (Ch. 5).

(continued on next page)

FIGURE 6-16. *(Continued)*

Charging circuits with regulator/rectifier

EXCEPT: • Tecumseh 7A
         • Syncro 20A
         • Others that do not tolerate battery or ac
           output–lead disconnects. Refer to text (Ch. 5).
PRECONDITIONS: Good, well-charged system battery.

```
┌────────────────────┐     ┌────────────────┐     ┌──────────────────────┐
│ Check battery voltage│────▶│  Voltage high  │────▶│ Partially discharge  │
│  with engine shut   │     │  12.6 - 13.5 V.│     │   battery with a     │
│       down.         │     └────────────────┘     │   controlled load    │
└────────────────────┘                             │  (e.g.,headlamps);   │
         │                                         │    need a reading    │
         ▼                                         │    of 12-12.5 V.     │
┌────────────────────┐     ┌────────────────┐      └──────────────────────┘
│   Voltage normal    │────▶│  Run engine near│◀─────────────┘
│     12-12.5 V.      │     │  governed speed │
└────────────────────┘     │  for 3-5 minutes.│
                           └────────────────┘
```

```
┌──────────────────┐   ┌──────────────────┐   ┌──────────────────┐
│  Voltage normal  │   │   Voltage low    │   │   Voltage high   │
│  13.6 - 14.7 V   │   │   12 - 12.5 V    │   │  14.8 V or higher │
│ with engine running│ │ with engine running│ │ with engine running│
└──────────────────┘   └──────────────────┘   └──────────────────┘
         │                      │                      │
         ▼                      ▼                      ▼
   ╭──────────╮          ┌──────────────┐       ┌──────────────┐
   │ System OK.│         │ Stop engine. │       │   Replace    │
   ╰──────────╯          │ Disconnect ac│       │regulator/rectifier│
                         │ output leads to│      │   (Ch. 5).   │
                         │regulator/rectifier.│   └──────────────┘
                         └──────────────┘
                                │
                                ▼
                  ┌──────────────────────────┐
                  │     Start engine,        │
                  │ cautiously increase speed│
                  │  while monitoring voltage│
                  │    across ac leads. Run  │
                  │   engine at **no more** than │
                  │   3/4 throttle or less to│
                  │    obtain 22-24 V.       │
                  │        (5-16E)           │
                  └──────────────────────────┘
```

```
┌──────────────────┐        ┌──────────────────┐
│ ac voltage normal │        │ No or low voltage.│
│    22-24 V.       │        └──────────────────┘
└──────────────────┘                 │
         │                           ▼
         ▼                   ┌──────────────────┐
┌──────────────────┐         │  Replace stator  │
│ Replace defective │        │    (Ch. 5).      │
│regulator/rectifier│        └──────────────────┘
│    (Ch. 5).       │
└──────────────────┘
```

FIGURE 6-16. *(Continued)*

Caution: On any system you are not familiar with, contact a factory-trained mechanic or a manufacturer's tech rep before making hot disconnects.

The two known mavericks are the Tecumseh-supplied 7A system and the Synchro 20A system with separate regulator and rectifier. The Tecumseh 7A system, found on some 3- to 10-hp side valve engines and the overhead valve OHV 120, uses any of three under-shroud regulator-rectifiers shown in Figure 6-17; the caption describes the ac voltage test procedure. Synchro regulators and rectifiers are clearly labeled with their manufacturer's name. Bring these systems to a dealer for service.

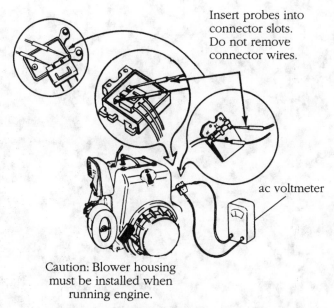

Insert probes into connector slots. Do not remove connector wires.

ac voltmeter

Caution: Blower housing must be installed when running engine.

FIGURE 6-17. *Tecumseh 7A system cannot tolerate open-circuit ac voltage measurements. Test ac output as shown, with regulator/rectifier electrically connected and the engine cooling shroud in place. (Unlike other rectifier/ regulators, these units do not require an engine ground.) Minimum acceptable stator performance is: 16 VDC at 2500 rpm; 19 VAC at 3000 rpm; 21 VAC at 3300 rpm.*

7

Engine mechanical

Chapter 2 describes quick checks for compression, bearing side play, and crankshaft straightness that should be made before an engine is torn down. Table 7-1 lists the more common two- and four-cycle maladies, together with their probable causes. Figure 7-1 describes why four-cycle engines develop a thirst for oil, and Figure 7-2 explains where the power goes.

Repairs to two-stroke engines are generally confined to piston-ring and crankshaft-seal replacement. More comprehensive repairs are impractical, at least for discount-house throwaways (Fig. 7-3).

Cylinder head

Four-cycle engines employ demountable cylinder heads sealed with composition gaskets and secured by cap screws.

Warning: When dealing with vintage engines, treat the gaskets as toxic material. Asbestos was phased out during the 1970s, but replacement gaskets may date from the time when this material was used.

Remove carbon deposits from the combustion chamber with a dull knife and a wire wheel. Try not to gouge the aluminum, especially the gasket surfaces.

Check head distortion with the aid of a piece of plate glass. If a 0.003-in. feeler gauge can be inserted between the bolt holes, the gasket surface should be refinished (Fig. 7-4). Tape a sheet of medium-grit wet-or-dry emery paper to the plate glass and, applying pressure to the center of the casting, grind in a figure-8 pattern. Oil makes the work go faster. Stop when the surface takes on a uniform sheen.

TABLE 7-1. Engine-related malfunctions
(Assuming ignition, fuel, and starting systems are functional)

Symptoms	Probable causes
Crankshaft locked	Jammed starter drive
	Hydraulic lock—oil or raw fuel in chamber
	Rust-bound rings (cast-iron bores only)
	Bent crankshaft
	Parted connecting rod
	Broken camshaft
Crankshaft drags when turned by hand	Bent crankshaft
	Lubrication failure, associated with cylinder bore and/or connecting rod
Crankshaft alternately binds and releases during cranking (rewind or electric starter)	Bent crankshaft
	Incorrect valve timing
	Loose blade/blade adapter (rotary lawnmower)
	Loose, misaligned flywheel
No or weak cylinder compression	Blown head gasket
	Leaking valves
	Worn cylinder bore/piston/piston rings
	Broken rings
	Holed piston
	Parted connecting rod
	Incorrect valve timing
No or imperceptible crankcase compression (two-cycle)	Leaking crankcase seals
	Leaking crankcase gaskets
	Failed reed valve (engines so equipped)
Rough, erratic idle	Stuck breather valve
	Leaking valves

Symptoms	Probable causes
Misfire, stumble under load	Improper valve clearance Weak valve springs Leaking carburetor flange gasket Leaking crankcase seals (two-cycle)
Loss of power	Loss of compression Leaking valves Incorrect valve timing Restricted exhaust ports/muffler (two-cycle) Leaking crankcase seals (two-cycle)
Excessive oil consumption (four-cycle)	Faulty breather Worn valve guides Worn or glazed cylinder bore Worn piston rings/ring grooves Worn piston/cylinder bore Clogged oil-drain holes in piston Leaking oil seals
Engine knocks	Carbon buildup in combustion chamber Loose or worn connecting rod Loose flywheel Worn cylinder bore/piston Worn main bearings Worn piston pin Excessive crankshaft endplay Excessive camshaft endplay Piston reversed (engines with offset piston pins) Loose PTO adapter
Excessive vibration	Loose or broken engine mounts Bent crankshaft

(continued on next page)

TABLE 7-1. *(Continued)*

Symptoms	Probable causes
Oil leaks at crank-shaft seals	Hardened or worn seals
	Scored crankshaft
	Bent crankshaft
	Worn main bearings
	Scored oil seal bore allowing oil to leak around seal outside diameter
	Seal tilted in bore
	Seal seated too deeply in bore blocking oil return hole
	Breather valve stuck closed
Crankcase breather passes oil (four-cycle)	Leaking gasket
	Dirty or failed breather
	Clogged drain hole in breather box
	Piston ring gaps aligned
	Leaking crankshaft oil seals
	Valve cover gasket leaking (overhead valve engines)
	Worn rings/cylinder/piston

Install the new gasket, lubricate the bolts with 30-weight motor oil, and torque in three equal increments to specification. Four-bolt heads tighten in an X pattern. Others are torqued as one would iron a shirt, that is from the center outward. As shown in Figure 7-5, this general rule does not always hold. Consult the factory manual for the engine in question.

Valves

Either of the spring compressors shown in Figure 7-6 can be used to extract and install side (block-mounted) valves. Rotate the crankshaft to seat the valves, insert the tool under the valve collar, compress the spring, and withdraw the locks.

Lock installation goes easier with a magnetic insertion tool, such as a Snap-on CF 771. When properly seated, the locks are swallowed by the collar and no longer visible.

Valve guides —
Must be within toler-
ances to prevent oil
from entering com-
bustion chamber and
exhaust gases from
entering crankcase.

Cylinder wall finish — Cylinder
wall glaze must be broken prior
to installing new piston rings to
allow rings to seat and control
oil. The desired finish also acts
as reservoir for oil to lubricate
rings and piston.

Piston rings —
Are to fit square-
ly to cylinder
wall with proper
end gap and ring
to groove
clearance, with
inside chamfer
to top of piston.

Breathers — Must
operate properly
to prevent oil
from being
expelled out
of engine.

Piston fit — Must
be within specs
for proper oil
control.

Drain holes —In
breather box and
under oil seals
must be clear to al-
low oil to return to
crankcase.

Oil passages —
Must be clear for
proper oil distribu-
tion to load-carrying
bearing surfaces.

Piston pas-
sages—Must be
clear to allow oil
to return to
crankcase.

Bearings — Should
be to spec to pre-
vent excessive oil
spillage and cause pressure
loss in pressure systems.

Gasket surfaces — Must be
clean and smooth. Use new
gaskets.

Oil filler caps — Must always
be tight and gasketed to prevent spillage out of breather.

Oil level — Overfilling will
cause leaking, burning and
oil-fouled spark plugs.

Engine speed — High speeds
will cause excessive oil consumption by burning and leaking.

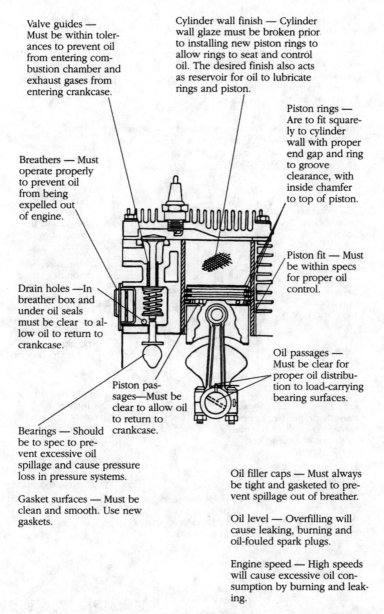

FIGURE 7-1. *Factors that affect oil consumption (four-cycle engines).*

Ignition — Must be properly timed so that spark plug fires at precise moment for full power.

Cylinder head — Should not be warped. Gasket surface must be true.

Valves — Check for seating, warping, sticking. Grind and lap to proper angle.

Cylinder head bolts — Tighten to proper torque.

Valve seats — Must be of specified angle and width.

Spark plug gap — Adjust to proper setting, use round feeler guage.

Cylinder head gasket — Must form perfect seal between cylinder and head.

Valve guide — Examine for wear, varnish which may prevent proper valve action.

Fins — Keep clean to prevent power loss because of over-heating.

Valve springs — Check free length, must have proper tension to close valve and hold on seat.

Piston rings — Piston rings must be fitted properly with recommended end gap to ensure sufficient pressure on cylinder wall to transfer heat and seal high pressure.

Valve gaps — Must be adjusted properly.

Piston pin — Must allow friction free movement of connecting rod and piston.

Cam lobes — Check for wear, must be proper size to open valve fully to allow complete discharge of exhaust and intake of fuel.

Oil passages — All oil holes and passages must be clear to allow full lubrication for friction free operation.

Connecting rod — Match marks must be matched and connecting rod nuts tightened to proper torque.

Piston fit — Must be fitted to cylinder with recommended clearance.

Air filter — Should be clean to allow engine to breath.

Carburetor — Must be set properly to assure proper and sufficient air and fuel.

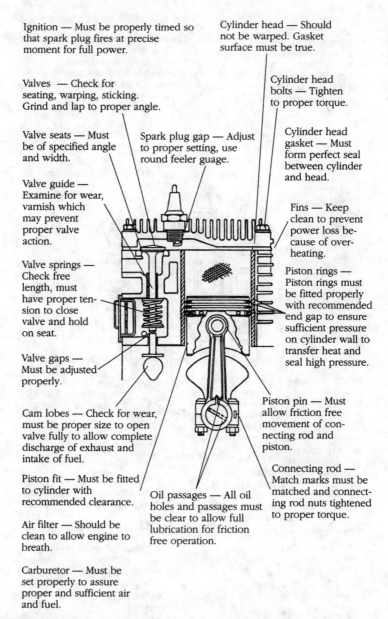

FIGURE 7-2. *Factors that affect power output (four-cycle engines).*

FIGURE 7-3. *The Ryobi in the lower part of the photo was repaired with a connecting rod and piston cannibalized from the second engine. New parts would have been prohibitive.* Photo by Tony Shelby

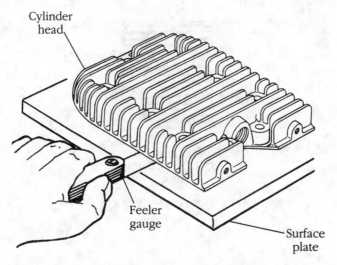

FIGURE 7-4. *Cylinder head flatness should be checked to assure gasket integrity. A piece of plate glass can be substituted for the surface plate shown.*

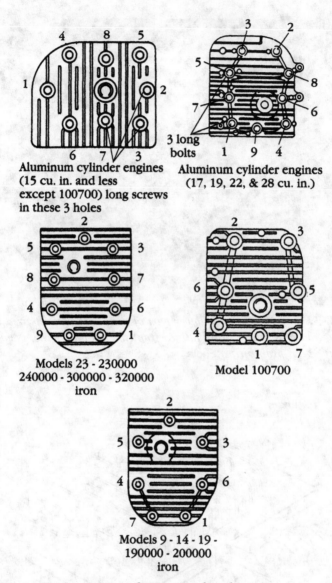

4 8 5
1   2
6 7 3

Aluminum cylinder engines
(15 cu. in. and less
except 100700) long screws
in these 3 holes

3   2
5   8
7   6
3 long
bolts 1 9 4

Aluminum cylinder engines
(17, 19, 22, & 28 cu. in.)

2
5   3
8   7
4   6
9   1

Models 23 - 230000
240000 - 300000 - 320000
iron

2
5   3
6   5
4
1   7

Model 100700

2
5   3
4   6
7   1

Models 9 - 14 - 19 -
190000 - 200000
iron

FIGURE 7-5. *Torque sequence for Briggs side-valve engines.*

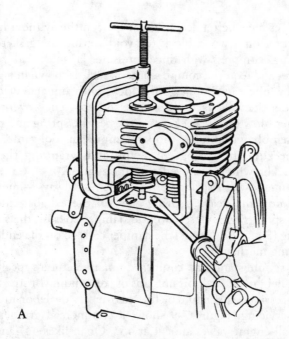

A

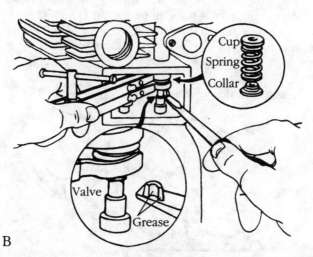

B

FIGURE 7-6. *Use a clamp (A) or bridge-type (B) spring compressor to remove and install valves mounted in the block. The former tool is available from Kohler and the latter from Briggs & Stratton. Note how split valve locks are spooned into place with a grease-coated screwdriver.*

Overhead valves have better accessibility. Detach the cylinder head and support the casting on a wood block to avoid scaring the gasket surface. Some valve springs compress with finger pressure. Japanese and Tecumseh springs are stouter and require a compressor tool, such as shown in Figure 7-7. In an emergency, one can disengage the locks with impact. Place a soft wood block under the valve face and a large socket wrench over the valve collar. A hammer blow on the socket compresses the spring and pops the keepers off. Assembly without the proper compressor tool can be accomplished by squeezing the springs in a vise and wrapping them with fine-gauge wire. The wire is retrieved after installation.

Exhaust and intake valve springs often—but not always—interchange. When they do not, the heavier spring goes on the exhaust side. Some engines employ springs with closely wound damper coils on the stationary side of the spring (Fig. 7-8). That is, the damper coils go on the end furthest from the actuating mechanism.

Valve springs should stand flat, conform to manufacturer's specifications for freestanding height, and exhibit no signs of coil binding or stress pitting.

Worn valve faces and seats should be turned over to a dealer or automotive machinist for servicing. Figure 7-9 shows a commercial grinder in use. While most small-engine valves are cut at 45°, Onan likes 44° and Briggs sometimes employs 30° on the intakes and 45° on the exhaust. Valve work goes nowhere without factory documentation (Fig. 7-10). Let the shop know that you have access to the documentation and will inspect their work to see that it conforms to it.

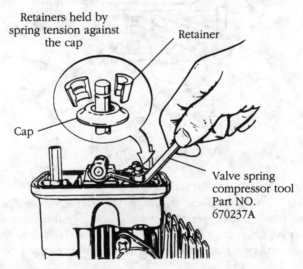

FIGURE 7-7. *Overhead-valve keepers can be released with a simple spring compressor or, as explained in the text, shocked loose.*

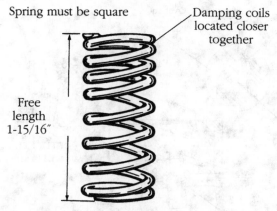

Spring must be square

Damping coils
located closer
together

Free
length
1-15/16″

For valve-in-head engines

FIGURE 7-8. *Install variable-rate springs with the damper coils on the stationary ends, furthest from the actuating mechanism. Springs should be replaced as part of every overhaul, especially on ohv engines where spring failure can result in a swallowed valve.*

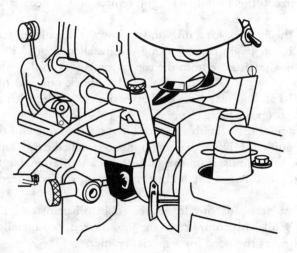

FIGURE 7-9. *A high-speed valve grinder.*

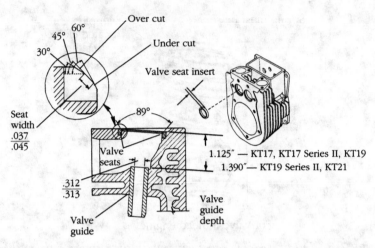

FIGURE 7-10. *An idea of the crucial nature of valve geometry can be seen from this Kohler-supplied drawing.*

Valve guides

As a rule, the wear limit for valve guides is 0.004 in., a figure difficult to detect without a set of plug gauges. If a valve exhibits perceptible wobble when wide open, the guide probably needs replacement. Some engine makers supply valves with oversized stems, so that the original guide can be reamed oversize. Others would have you replace and, if necessary, ream the guides.

Normally this is work for a machinist. However, a patient mechanic can usually pull it off. Begin by measuring the installed depth of the original guides, that is, the distance from the top of the guides to the valve seats, as called out in Figure 7-10. Side-valve guides knock out and install from above (Fig. 7-11). You may have to shatter the old guides with a punch to retrieve them from the valve chamber.

Fabricate a pilot driver with a reduced diameter on one end that exactly matches the guide ID. Drive the new guide to installed depth and test with a valve. If the valve binds, ream the guide 0.0015-0.0020 in. larger than the stem diameter. Because of the odd sizes involved, you will need to use an adjustable reamer.

Overhead-valve heads should be heated in oil prior to guide service (Fig. 7-12). While this complicates the job, the reduced installation force usually eliminates the need for a special reamer.

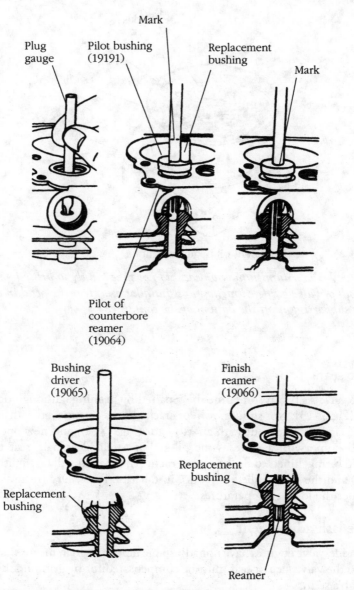

FIGURE 7-11. *Installation of valve guide bushings goes much easier if you have the correct tools.*

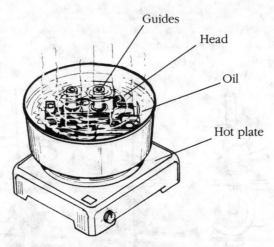

Guides

Head

Oil

Hot plate

Heat until oil begins to smoke

FIGURE 7-12. *Aluminum ohv heads do not take kindly to brute-force methods of valve guide extraction and installation. Heat the head in oil while supporting it off the bottom of the container.*

Valve seats

Loose valve seats can sometimes be repaired by peening, although don't bet on it (Fig. 7-13). Nor should worn or cracked seats be removed with a punch from below or pried out with an extractor (Fig. 7-14), and new ones hammered home with a valve as the pilot. For a reliable repair, seat recesses should be machined to fit the replacement part (which will be slightly oversized) and the seats chilled to reduce installation force. Dry ice and alcohol produce the lowest temperatures.

Valve lash adjustment

Most side-valve engines have nonadjustable tappets, which means that metal lost to the valve face or seat must be compensated for by grinding the tip of the valve stem.

Install the valve without the spring, turn the crankshaft until the valve rides on the heel of the cam, and measure the clearance with a feeler gauge (Fig. 7-15). Carefully grind the stem, just "kissing" the wheel, to obtain the specified clearance (typically 0.006-0.008 in. for the intake and 0.010-0.013 in. for the exhaust). Take off too much and the associated valve or seat will have to be reground. Finish by breaking the sharp edges with a stone.

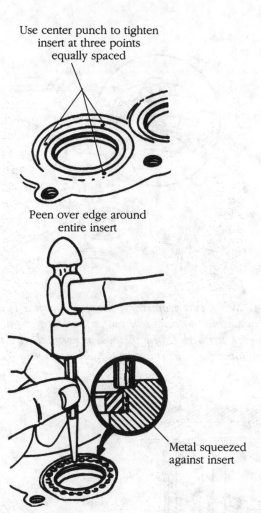

Use center punch to tighten
insert at three points
equally spaced

Peen over edge around
entire insert

Metal squeezed
against insert

FIGURE 7-13. *Some mechanics attempt to repair loose valve seats by peening. That rarely, if ever, works. Peening is also used as insurance when new seats are installed. A better approach is to have the machinist recess the new seat about 0.030 in. below the surrounding metal. Then, using a flat-tipped punch, roll the metal over the edge of the seat.*

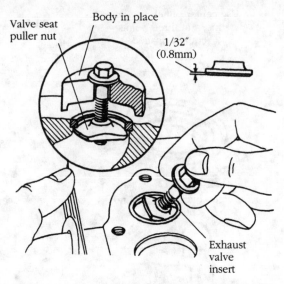

FIGURE 7-14. *When port geometry makes it impossible to drive the seats out from below with a punch, a seat puller fabricated from an old valve can be used. However, neither of these methods is optimal. To eliminate damage to the seat bores, have a machinist cut the old seats out.*

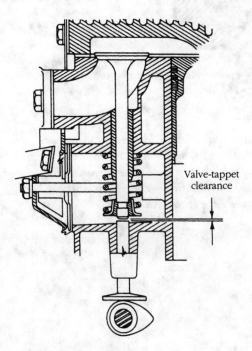

FIGURE 7-15. *Valve lash on side-valve engines is measured between the end of the valve stem and the tappet, with the tappet on the cam base circle.*

Some of the better side-valve and all overhead-valve engines have provisions for valve-lash adjustment. Lash for ohv engines appears as the clearance between the rocker arm and valve stem. Adjustment screws for engines with shaft-supported rocker arms bear against the pushrods (Fig. 7-16). Other ohv engines use stamped-steel rockers that pivot on fulcrum nuts. The fulcrum nuts, secured to their studs by setscrews or locknuts, control lash by varying the height of the rocker arms (Fig. 7-17).

To adjust lash on ohv engines, rotate the crankshaft to bring the associated tappet on the heel of the camshaft and loosen the locknut or screw. Move the adjustment screw or fulcrum nut as necessary to achieve the required clearance. Tighten the lock and check the adjustment, which will have drifted a few thousandths.

Push rods

Inspect push rods for wear caused by contact with the guide plates and for straightness. It is possible to salvage bent rods with judicious vise work, but the practice is an expedient, resorted to when replacement parts cannot be had.

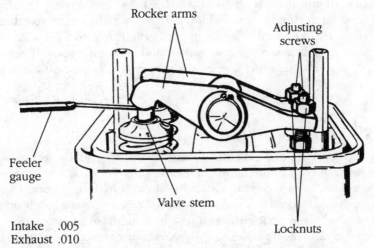

FIGURE 7-16. *Lash for overhead valves is measured between the end of the valve stem and rocker arm. Shaft-mounted rockers carry the adjustment screws.*

FIGURE 7-17. *Pressed-steel rockers pivot on adjustable fulcrum nuts, secured by set screws or lock nuts.*

Breathers

Four-cycle engines include a crankcase breather connected by a flexible line to the carburetor intake (Fig. 7-18). The filter element, in conjunction with a baffle, separates liquid oil from blow-by gases, which then recycle through the carburetor. The reed valve maintains a slight negative pressure in the crankcase to reduce seepage past gaskets and crankshaft seals. If the filter clogs or the valve becomes inoperative, the engine pumps oil like a mosquito fogger.

Reed valves

Many two-cycle engines use a functionally similar device in the form of a reed valve between the carburetor and crankcase (Fig. 7-19). The reed assembly acts as a check valve to contain the air-fuel mixture in the crankcase. Contact surfaces should be dead flat, and valve petals should either rest lightly on their seats or stand off by no more than a few thousandths of an inch.

Tecumseh also incorporates a reed-type compression release in some of its two-strokes. While these devices are rarely encountered, the engineering is worth showing (Fig. 7-20).

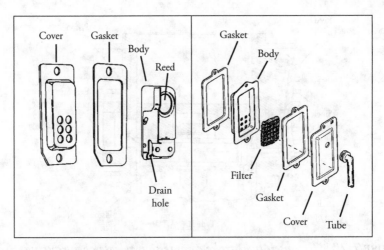

FIGURE 7-18. *Various Tecumseh crankcase breathers.*

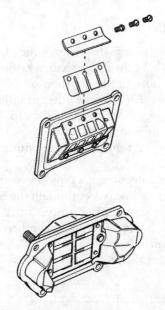

FIGURE 7-19. *Reed intake valves should stand off no more than 0.010 in. from their seats.* Tecumseh Products Co.

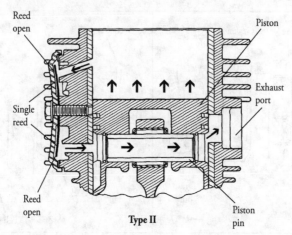

Reed
open

Piston

Single
reed

Exhaust
port

Reed
open

Type II

Piston
pin

FIGURE 7-20. *Tecumseh two-stroke compression release opens during cranking to bleed pressure through the piston pin and out the exhaust port.*

Pistons and rings

Disengage the flange or side cover as described in the caption to Figure 7-21. Two-piece connecting rods have their caps secured by bolts or studs. To avoid an assembly error, make note of the orientation of the rod-and-piston assembly relative to the camshaft or some other prominent feature. Loosen rod nuts in two or three steps, and remove the rod cap. Match marks on rod cap and shank must be aligned upon assembly. Drive the piston and attached rod shank out of the top of the bore with a hammer handle or wooden dowel (Fig. 7-22).

Figure 7-23 shows the architecture of a typical two-stroke engine. In this example, the crankcase parting line passes through the center of the cylinder bore. Each half of the crankcase carries a press-fitted ball- or roller-type main bearing and oil seal. String-trimmer and other inexpensive engines often get by with a single main bearing.

Note: Weed-eater centrifugal clutches may have left-hand threads.

Inspection

Bright, uniformly polished rings are the norm. Rings that stick in their grooves suggest poor maintenance (failure to change oil, dirty cooling fins) or abuse (lugging under load, insufficient power for the application). Broken rings result from improper installation or worn piston grooves.

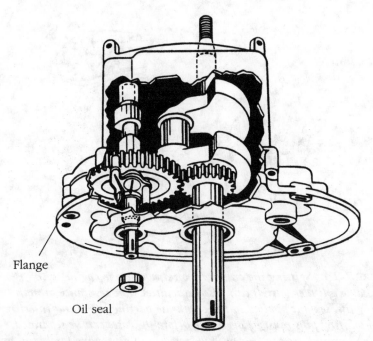

Flange

Oil seal

FIGURE 7-21. *The flange on vertical-crank engines locates the lower, or pto, main bearing. Before proceeding with disassembly, remove rust and tool marks from the crankshaft with an emery cloth and a file. Cover the keyways (which are sharp enough to cut the crankshaft seal) with a layer of Scotch tape. Lubricate the crankshaft and remove the flange hold-down cap screws. Position the engine on its side and separate the castings with a rubber mallet. The camshaft (on side- and overhead-head-valve engines) should remain engaged with the flywheel so that timing-mark alignment can be verified. Do not attempt to pry the flange off. The same general procedure holds for side covers on horizontal-shaft engines.*

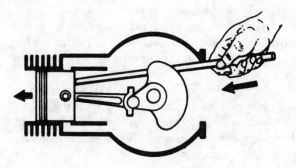

FIGURE 7-22. *Once the rod cap has been detached, use a wooden dowel to drive the piston assembly out.*

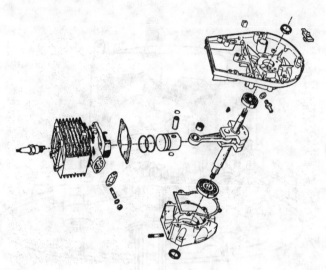

FIGURE 7-23. *Spitting the crankcase on a two-stroke engine with detachable cylinder barrels can pose difficulties. The resistance of main-bearing fits, sealant applied to the crankcase parting lines, and interference fits of locating pins must be overcome before the cases can be separated. Stubborn crankcases can be gently warmed with a propane torch and pried apart with a hammer handle inserted into the cylinder-barrel cavity. Exercise extreme care to avoid over-heating or warping the fragile castings.*

Examine the piston skirt for wear on the thrust faces at right angles to the piston pin. Figure 7-24 illustrates abnormal wear patterns produced by bent or twisted connecting rods. Forces that rock the piston can also drive the piston pin past its locks and into collision with the cylinder bore.

Deep scratches on the piston skirt result from chronic overheating that can leave splatters of aluminum welded to the bore. A dull, matted finish means that abrasives have been ingested, usually by way of a leaking air filter. Should this happen, hone the cylinder bore and replace the piston.

Pistons need about 0.002 to 0.003-in. bore clearance for thermal expansion, but wear limits are flexible. Lightly used four-strokes putter on for years with piston-to-bore clearances of 0.006 in. and more. High-revving two-strokes are less tolerant.

Pistons usually taper toward the crown to allow for expansion under thermal load. In addition, four-cycle pistons are cam ground with the thrust faces on the long axis. The piston remains centered in the bore when cold and expands to a full circle as temperatures increase. Two-stroke pistons are machined round to control leakage at startup.

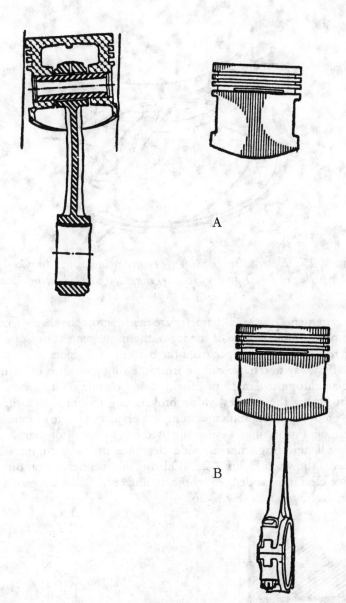

FIGURE 7-24. *As shown by the shaded lines in drawing A, a bent conn rod tilts the piston to create an hourglass-shaped wear pattern. A twisted rod rocks the piston, concentrating wear on the upper and lower edges of the skirt (drawing B).*

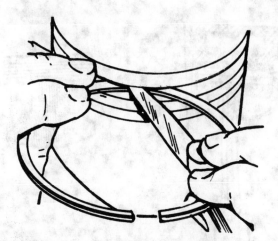

FIGURE 7-25. *Determine ring side clearance using a new ring as a gauge. The upper side of No. 1 groove (shown) takes the worst beating.* Onan

Measurements of piston diameter are made across the thrust faces at 90° to the piston pin. Because of the taper, the measurement must be made at the factory-specified distance from the bottom of the skirt.

The best way to remove carbon from the ring grooves is to farm out the job to an automotive machinist for chemical cleaning. Otherwise, you will need to scrape the grooves with a broken ring mounted in a file handle. (Ring-groove cleaning tools are, in my experience, a waste of money.)

Warning: Piston rings—especially used rings—are razor sharp.

Using a new ring, measure side clearance on both compression-ring grooves as shown in Figure 7-25. (Oil-ring grooves never wear out.) Excessive side clearance, as defined by the manufacturer, allows the ring to twist during stroke reversals (Fig. 7-26).

FIGURE 7-26. *A major cause of ring breakage is the twist created by worn ring grooves.*

Piston pins

Four-cycle piston-pin bearing wear is relatively uncommon because of the thrust reversals every second revolution. Compression and expansion strokes bear down on the pin, exhaust and intake strokes drive the pin from below. Two-stroke pins are subject to an almost constant downward thrust that tends to squeeze out the lubricant. In either case, the bearing is considered acceptable if it has no perceptible up-and-down play and if the piston pivots on the pin from its own weight.

Most pistons incorporate a small offset relative to their pins. Consequently, one must install the piston as found. An arrow or other symbol on the crown marks leading edge or indexes with some other reference such as the flywheel.

Remove and discard the locks. New circlips are inexpensive insurance against the pin drifting into contact with the cylinder bore. If the piston and rod assembly are out of the engine, drive or press the pin out, being careful not to score the pin bores. When the connecting rod remains attached to the crankshaft, extract the pin with the tool shown in Figure 7-27 or heat the piston. The safest, and surely the messiest way, to apply heat is to wrap the piston with a rag soaked in hot oil.

Installation is the reverse process. Lubricate the pin and pin bosses with motor oil or assembly lube. Make certain that the pin locks seat around their whole circumferences.

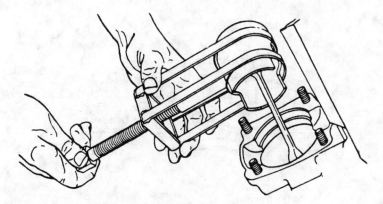

FIGURE 7-27. *A piston-pin extractor can be ordered through motorcycle and snowmobile dealers. A Kohler tool is shown.*

Piston rings

Four-cycle engines usually have three rings. Counting from the top, we have No. 1 compression ring, No. 2 compression (or scraper) ring, and the oil ring. The latter may be cast in one piece or made up of segments. Two-stroke engines are fitted with two identical compression rings, usually fixed in their grooves by pegs. (Were the rings free to rotate, the ends could snag on the ports.)

Careful mechanics check the end gap of each ring. Using the piston crown as a pilot to hold the ring square, insert the ring about midway into the cylinder (Fig. 7-28). Measure the gap with a feeler gauge.

Most manufacturers call for about 0.0015 in. of ring gap per inch of cylinder diameter. Too large a gap suggests that the bore is worn or that the ring is undersized for the application. Too small a gap leads to rapid cylinder wear and shattered rings. Correct by filing the ends square.

Installation

Lay out rings in the order of installation. Make certain that you have correctly identified each ring and each ring's upper side, which should be marked (Fig. 7-29). The lowest ring goes on first. Using the expander shown in Figure 7-30, spread the ring just enough to slip over the top of the piston and deposit it into its groove. Verify that rings seat into their grooves and that ring ends of two-stroke pistons straddle their pegs.

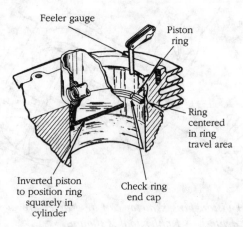

Feeler gauge

Piston ring

Ring centered in ring travel area

Inverted piston to position ring squarely in cylinder

Check ring end cap

FIGURE 7-28. *Using the piston as a pilot, insert each replacement ring about halfway into the bore and measure its gap. Variations in gap as the ring moves deeper into the bore give some idea of cylinder taper.*

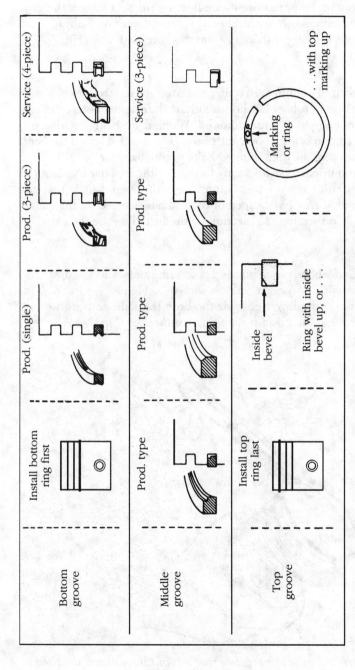

FIGURE 7-29. *Kohler ring sequence and orientation is typical of four-cycle engines.*

In order to contain compression, rotate floating rings to stagger the gaps 120°. On Tecumseh engines with relieved valves (shades of flathead Ford hotrods!), position the ring ends away from the bore undercut (Fig. 7-31).

Integral barrel

Bring the crankshaft to bottom dead center and cover the rod studs (when present) with short pieces of rubber fuel line. Lubricate the cylinder bore, crankpin, rod bearings, pin, and piston with motor oil. Without upsetting the ring-gap stagger, install a compressor tool over the piston (Fig. 7-32). Tighten the band only enough to squeeze the rings flush with the piston diameter.

With the piston oriented as originally found, set the tool hard against the fire deck, and carefully tap the piston into the bore. Do not force matters. If the piston binds, a ring has escaped or the rod has snagged on the crankshaft. Read the "Connecting rod" section before installing the rod cap.

Detachable barrel

Lubricate the cylinder bore, piston pin, and piston ring areas. Support the piston on the crankcase with a wooden fork, as shown in Figures 7-33 and 7-34. Most factories bevel the lower edge of the bore to facilitate piston entry. Straight-cut bores call for a clamp-type ring compressor (Fig. 7-33).

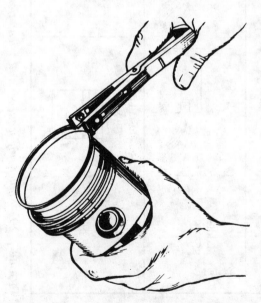

FIGURE 7-30. *Installing a compression ring on an Onan piston with the aid of a ring expander. Expand the rings only enough to slip them over the piston.*

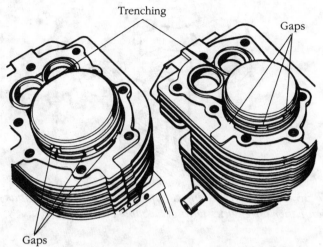

Stagger ring end gaps away from trenching

FIGURE 7-31. *Rings for Tecumseh engines with trenched valves install with their ends turned away from the undercut.*

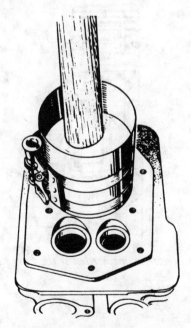

FIGURE 7-32. *A ring compressor sized for small engines is used when the piston installs from the fire deck. Use a hammer handle to gently tap the piston home.*

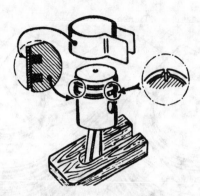

FIGURE 7-33. *A homemade ring clamp and a wooden block make ring installation easier for engines with detachable cylinder barrels.*

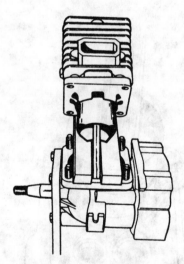

FIGURE 7-34. *A compressor is not needed if the bore has a taper on the lower edge.*

Cylinder bores

All discount-house and a handful of upper-echelon engines, such as the Briggs 11-CID Intec, run their pistons directly against the aluminum block metal. While soft metal cylinders can be rebored to accept oversized pistons (chrome-plated to reduce scuffing), the exercise seems futile. Aluminum-bore engines have a working life of 200 hours or so. It is possible to upgrade these engines with cast-iron cylinder liners. Expect to pay $80 to $150 for this service. In the discussion that follows, I assume you are working with cast iron.

Examine the bore for deep scratches, aluminum splatter from piston melt, and for the cat's tongue texture that comes from silicon particles ingested past a faulty air cleaner. Maximum wear occurs just below the upper limit of ring travel, where heat is greatest, lubrication minimal, and corrosives most concentrated. In the past, upper cylinder wear could undercut the bore enough to leave a ridge. Thanks to modern lubricants, the ridge and its corrective, the ridge reamer, have pretty well passed into history.

It was once considered necessary to roughen the cylinder with a hone to seat new rings. Some manufacturers continue to insist on honing; others say that wear, however induced, is wear. The decision is up to the rebuilder.

Boring cylinders is a job that should be relegated to an automotive machinist who has the tools and set-up expertise to bore at 90° to the crankshaft centerline.

Confusion arises because Briggs and a few other small-engine makers size replacement pistons relative to the diameter of the original piston. A Briggs piston stamped ".030" measures thirty thousandths of an inch larger than the standard piston. If the bore is machined 0.030-in. over its original diameter, the replacement piston will have the correct running clearance. Automotive practice is to base piston oversizes on the bore, which means that a piston marked ".030" is 0.030 in. larger than the original bore diameter. For the piston to have room for expansion, the bore must be machined to 0.032 or 0.033 in.

Connecting rods

Cast or forged aluminum is the material of choice for four-cycle rods. Utility and light-duty engines run their crankpins directly against rod metal. Figure 7-35 illustrates this type of construction. Some manufacturers supply undersized rods so that the crankpin can be reground.

Better quality rods have precision bearing inserts at the big end and a brass or bronze bushing at the small end (Fig. 7-36). Under-sized inserts (0.010 and 0.020 in. for American-made engines) permit the crank to be reground.

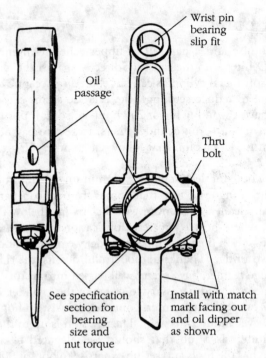

Wrist pin
bearing
slip fit

Oil
passage

Thru
bolt

See specification
section for
bearing
size and
nut torque

Install with match
mark facing out
and oil dipper
as shown

FIGURE 7-35. *Aluminum is a favorite material for small-engine rods.*

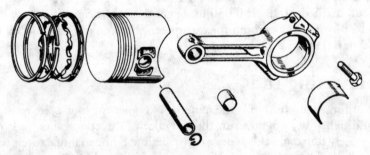

FIGURE 7-36. *Onan connecting rods feature a brass-bushed small end and precision inserts at the big end.*

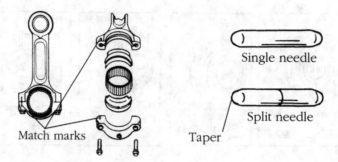

FIGURE 7-37. *Two-stroke rod with needle bearings at the big end and a brass bushing at the eye.*

Figure 7-37 shows an aluminum rod for a two-cycle engine with single or split needle bearings that run on steel races at the big end, and a bushing at the rod eye. String trimmers and the like use stamped-steel rods, which cost less than aluminum.

Catastrophic rod failure usually involves the big-end bearings. Plain bearings skate on a wedge of oil that develops soon after startup. Once up to speed, the bearing makes no contact with its journal. Insufficient bearing clearance prevents the oil wedge from forming; too much clearance causes the wedge to leak down faster than it can be replenished. In either case, the result is metal-to-metal contact, fusion, and a thrown rod.

Needle bearings make rolling contact against their races without the cushion of an oil wedge. Consequently, any discontinuity—fatigue flaking, a spot of rust, skid marks—results in bearing seizure.

Incorrect assembly can also result in rod breakage. Big ends crumble into bite-sized chucks when the fasteners have insufficient preload. Proper torque might not have been applied during assembly or the rod locks might have failed, allowing the bolts to vibrate loose. This is why manufacturer's torque specifications have the authority of Holy Writ and why new locks or lock nuts should be installed whenever a connecting rod is disassembled. Bend-over tab locks usually carry a spare tab that can be used during the first overhaul. Once a tab has been engaged, it should not be straightened and reused.

Rod orientation

Correct orientation has three components:

- **Piston-to-rod.** As mentioned earlier, the piston pin may be offset relative to the piston centerline. Wrong assembly results in knocking.

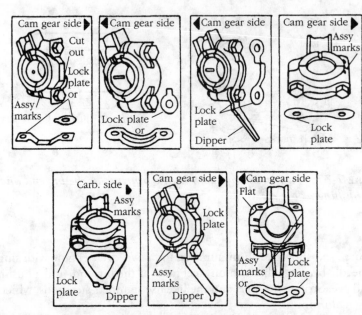

FIGURE 7-38. *Briggs & Stratton rod-to-engine and cap-to-rod orientation. McCulloch and a few other manufacturers fracture their rod caps after machining. When assembled correctly, the parting line becomes almost invisible.*

- Rod-to-engine. Some connecting rods are drilled for oil transfer; others are configured so that reverse installation locks the crankshaft.
- Cap-to-rod shank. In order to maintain precision, rods and caps are assembled at the factory and reamed or diamond-bored to size. Stamped or embossed marks identify cap orientation (Fig. 7-38). Failure to assemble the cap correctly results in early and catastrophic failure.

Rod inspection

The piston should pivot of its own weight on the rod eye when held at 45° off the vertical at room temperature. Pin-to-piston fits are tighter, but loosen when the piston reaches operating temperature. In no case should the piston wobble or tilt on its pin. Replacement rod-eye bushings sometimes install without the need for finish reaming, but do not count on it.

The big-end bearing is the most critical rubbing surface and never more so that when the bearing consists of needles or rollers. Any discontinuity on the crank pin means that both the crankshaft and rod bearings must be replaced if the engine is to live. A single rust pit sets in motion a chain of events that culminates in rod seizure.

Measure the crank pin at several places along its length and around its circumference with a good-quality micrometer—dial or electronic calipers do not have the requisite precision. Taper and out-of-round should be 0.001 in. or less. Do the same for plain-bearing connecting rods. The difference between rod ID and crankpin OD is the running clearance, which should be no more than 0.0030 in.

Caution: Do not attempt to restore bearing clearances by filing the rod cap.

That said, most mechanics approximate main-bearing clearances with plastic-gauge wire, available in various thicknesses from auto parts stores. Sealed Power SPG-1 Plastigage comes in three color-coded sizes—green reports a clearance range of 0.002 in. to 0.003 in., red spans 0.002 in. to 0.006 in., and blue 0.004 in. to 0.009 in. Everyone, even those who have access to precision measuring instruments, should use the wire as an assembly check.

Follow this procedure:

1. Turn the crankshaft to bring the rod to bottom dead center.
2. Remove the rod cap.
3. Wipe off any oil on the rod and crankpin.
4. Tear off a piece of green gauge wire and lay it along the full length of the journal (A in Fig. 7-39).
5. Install the rod cap, with match marks aligned, and torque down evenly to factory specs. The crankshaft must remain stationary as the bolts are tightened.
6. Remove the cap and compare the width of the wire against the scale printed on the envelope (B in Fig. 7-39). Average width corresponds to bearing clearance; variations in width indicate the amount of crankpin taper.
7. Repeat the process with two pieces of gauge wire across the journal (C in Fig. 7-39). The relative widths of the wires are a crosscheck on taper and say something about out-of-round.
8. Scrape off all traces of the wire from the bearing and journal.

Caution: As gauge wire ages, it hardens and becomes less accurate. Shelf life is said to be about six months.

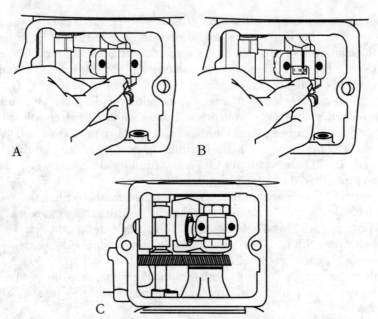

FIGURE 7-39. *Lay a piece of plastic gauge wire along the length of the crankpin (A). Assemble the cap and, without moving the crankshaft, torque the rod nuts to factory specs. Lift the cap off and measure the flattened wire against the scale on the package (B). Repeat the operation, positioning the wire at two points on the crankpin circumference to detect taper and out-of-round (C).*

Rod assembly

Coat all bearing surfaces with clean motor oil or assembly lube. Grease loose needle bearings to hold them in position around the periphery of the crankpin as the rod is installed (Fig. 7-40). All needles are accounted for if they pack closely around the crankpin with no space for another.

Check piston-to-block and piston-to-rod orientation one final time. Turn the crank to bottom dead center and guide the rod assembly home. Install the cap and verify its orientation. Tighten the rod bolts or studs evenly in three increments to specified torque.

Turn the engine over by hand to detect possible binds. The rod should move easily from side-to-side along the crankpin. Manufacturers do not often provide side-play specifications, but connecting rods need several thousandths of an inch of axial freedom.

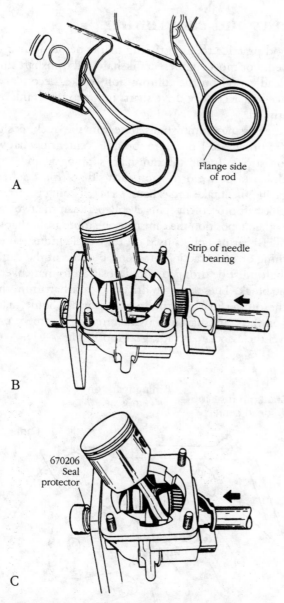

FIGURE 7-40. *TVS and TVXL840 rods present a special case. The rod installs with the flange toward the pto side of the engine (A). Grease-packed bearings go on the crankpin (B) and the rod is gingerly slipped over the crankpin and bearings (C). Note the factory-supplied seal protector, which is one of two needed for this job.*

Crankshafts and cam timing

It is always good practice to align timing marks before four-cycle engines are disassembled. For most engines, crankshaft and camshaft timing marks index at top dead center on the compression stroke. Secondary marks on rotating-balance and accessory-drive shafts index to the crank or cam after the valves are timed.

If the marks are missing or ambiguous, time from the "rock" position. Rotate the crankshaft to bring No. 1 piston to top dead center on what will become the compression stroke. Install the camshaft, slipping it under the tappets. Rock the crankshaft a degree or two on each side of top dead center (tdc), alternately engaging the intake and exhaust valves. Timing is correct when the free play in crankshaft movement splits evenly between the two valves. If one valve leads the other, reposition the camshaft one tooth from that valve.

Ball- and roller-bearing cranks used on four-cycle industrial engines can present something of an extraction problem. Because of the confined quarters, the camshaft must be dropped out of position to maneuver the crankshaft out of the block. These engines drive the camshaft from the magneto side or hide the timing mark behind a ball bearing. Some manufacturers stamp a mark on the crankshaft web (which makes alignment problematic) and others bevel the associated crank-gear tooth (Fig. 7-41).

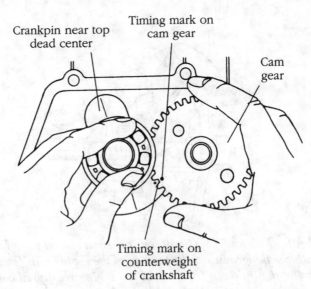

FIGURE 7-41. *Timing marks are not always visible at the point of tooth contact.* Briggs & Stratton Corp.

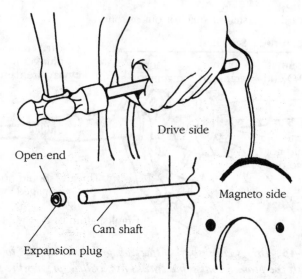

Drive side

Open end

Magneto side

Cam shaft

Expansion plug

FIGURE 7-42. *The classic Briggs and Kohler camshaft axle drives out through an expansion plug, which should be coated with a sealant before the cam is installed.*

Release the camshaft by driving out the cam axle, as shown for a Briggs engine in Figure 7-42. Classic Kohlers follow the same pattern. Note the expansion plug is coated with sealant before assembly. Timing goes easier if you color the associated pair of crankshaft teeth with a Magic Marker.

Figure 7-43 illustrates crankshaft inspection points. Give special attention to the crankpin, as described earlier. When the crank is drilled for pressure lubrication, it is good practice to remove the expansion plugs and clean the oil passages, which act as sludge traps.

Lightly polish the journals with No. 600 wet-or-dry emery paper saturated in oil. To avoid creating flat spots, cut a strip of emery paper as wide as the journal. Wrap the strip around the journal, and spin it with shoelace or leather thong.

Straightening crankshafts is a touchy subject fraught with legal complications for the mechanic who gets someone hurt. Even so, experienced and patient craftsmen routinely straighten cranks bent a few thousandths. If you want to pursue this matter, recognize that you are on your own.

Warning: No small-engine manufacturer recommends that crankshafts be straightened.

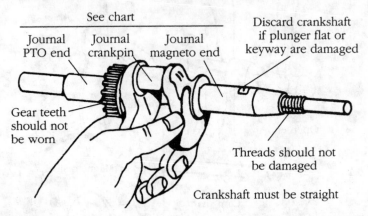

Discard crankshaft if small or out of round

See chart

Journal PTO end   Journal crankpin   Journal magneto end

Discard crankshaft if plunger flat or keyway are damaged

Gear teeth should not be worn

Threads should not be damaged

Crankshaft must be straight

FIGURE 7-43. *Briggs crankshaft inspection procedure applies to other makes, with the proviso that some cranks are drilled for pressure lubrication. Clean the oil passages and lightly chamfer the oil ports.*

The work requires two machinist's V-blocks, a pair of dial indicators, and a straightening fixture usually built around a hydraulic ram. Tremendous forces are involved. The crank is supported on the blocks at the main bearing journals with the indicators positioned near the ends of the shaft. Total run out should be no more than +/−0.001 in. (0.002 in. indicated). Using the fixture, bring the crank into tolerance in small increments with frequent checks. Once the indicators agree, send the shaft out for magnetic-particle inspection. Skipping this final step, which only costs a few dollars at an automotive machine shop, can be disastrous for all concerned. Crankshafts break, especially when bent and straightened.

Upon assembly, check endplay, or float. This check is made internally (Fig. 7-44) with a feeler gauge or from outside the engine with a dial indicator. The amount of float is not crucial so long as the shaft has room to expand. Specs fall in the 0.004 to 0.009 in. range. A doubled up or thicker flange/side-cover gasket increases float when the dimension has been reduced by a new crank or flange casting. A thrust washer—usually placed between the crank and pto main bearing and, occasionally, on the magneto side—compensates for wear.

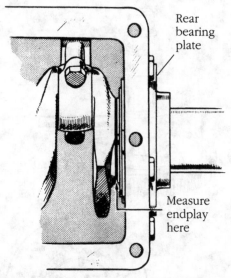

Rear
bearing
plate

Measure
endplay
here

FIGURE 7-44. *Crankshaft float measured with a feeler gauge on an Onan engine. A suitably mounted dial indicator may also be used.*

Camshafts

The camshaft lives in the block on side- and overhead-valve engines. As shown in Figure 7-42, the cams for Kohler Magnum and vintage Briggs engines ride on a steel pin. Other engines run their cams on plain bearings in the block and cover.

Cam failure of the kind that most mechanics flag is obvious: once the surface hardness goes on iron cams, the lobes rapidly wear round. Gear teeth occasionally fatigue and break. Interestingly enough, Briggs reports that its plastic cams generate fewer warranty claims than the metal versions.

Many camshafts include a compression release to aid starting (Fig. 7-45). These systems employ a pin or other protrusion that lifts one of the valves during cranking. When the "bumper" wears, the factory fix is to replace the camshaft. However, a welder can usually build up the worn surface with hard facing.

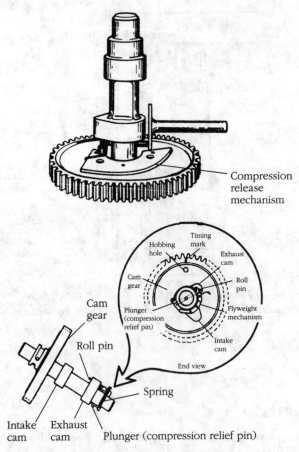

FIGURE 7-45. *Compression releases come in three types. Briggs Eazy-Spin employs a ramp ground on the cam profile that unseats the intake valve. These units give no problems. Others employ a bumper, either spring-loaded as shown or actuated by a linkage from the starter, to unseat one of the valves during cranking. The bumper is the weak spot.* Tecumseh Products Co.

Main bearings

The crankshaft runs on plain or anti-friction (ball or roller) bearings, or a combination of both types.

Antifriction bearings

Figure 7-46 shows a typical setup using two tapered roller bearings with provision for a hardened washer at the magneto side to control endplay. Check the condition of the bearings by removing all traces of lubricant and spinning the outer races by hand. Roughness or a tumbler-like noise means that the bearings have reached the end of their service lives.

Caution: Do not spin anti-friction bearings with compressed air.

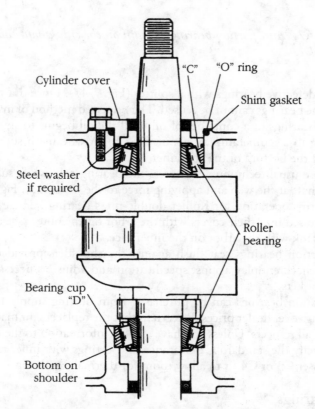

FIGURE 7-46. *Better engines use tapered roller bearings (as opposed to balls) that absorb large amounts of thrust as well as radial loads.*

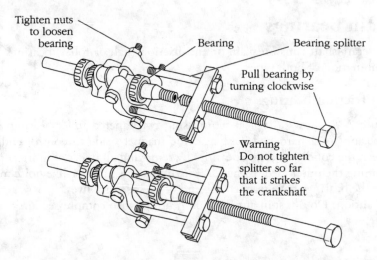

Tighten nuts
to loosen
bearing

Bearing

Bearing splitter

Pull bearing by
turning clockwise

Warning
Do not tighten
splitter so far
that it strikes
the crankshaft

FIGURE 7-47. *Anti-friction bearings remain on the crankshaft unless they will be replaced.*

Extract defective bearings with a splitter (Fig. 7-47). Once drawn in this manner, the bearings cannot be reused. The preferred method of installation is to heat bearings in a container of oil until the oil begins to smoke (corresponding to a temperature of about 475° F). A wire mesh supports the bearing off the bottom of the container.

The more usual technique is to press the bearing cold while supporting the crankshaft at the web and applying force to the inner race. Figure 7-48 illustrates this operation for Kohler double-press bearings. These bearings are first pressed into their covers with the arbor on the outer race and then over the crank with the arbor on the inner race.

Anti-friction bearings seat flush against the shoulders provided. Upon assembly, check endplay against specification and adjust as necessary with shims or gaskets.

Anti-friction bearings can be purchased from bearing supply houses at some savings over dealer prices. Be certain that the replacement matches the original in all respects. Unless you have reliable information to the contrary, do not specify the standard C1 clearance for bearings with inner races. Ask for the looser C3 or C4 fit to allow room for thermal expansion.

Plain bearings

In a perfect world, one would establish main-bearing clearances with inside and outside micrometers as described for crankpin bearings. That said, I have yet to see a small-engine mechanic do more than wobble the crankshaft. Most plain bearings are set up with 0.002-in. clearance new and tolerate something like twice that before the seals wear out.

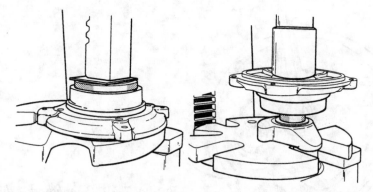

FIGURE 7-48. *Some Kohler engines use pto-side bearings with a double interference fit. The cup, or outer race, presses into the bearing cover and the inner race, together with the bearing and cover, presses over the crankshaft. A support under the crankshaft web nearest to the arbor isolates the crankpin from bending loads.*

Engines from major manufacturers can be rebushed, but the work is best left to a dealer who has the proper reamers and pilots. Shops that cater to racers can install Briggs DU™ Teflon-impregnated bronze bushings that withstand twice the radial loads of aluminum-block metal bearings.

Thrust bearings

Most manufacturers install a hardened steel washer between the flange/side cover and crankshaft cheek. Kohler and a few others specify proper Babbitt or roller thrust bearings. Poorly maintained vertical-shaft engines develop severe galling at the flange thrust face, which can be corrected by resurfacing or replacing the casting.

Seals

Seals, mounted outboard of the main bearings, contain the oil supply on four-cycle engines and hold crankcase pressure on two-strokes. Seal failure is signaled by oil leaks at the crankshaft exit points or, on two-cycle engines, by hard starting and chronically lean mixtures.

The old seals pry out with a flat-bladed screwdriver (Fig. 7-49). Install the replacement with the maker's mark visible and the steep side of the elastomer lip toward the pressure. Lubricate the lip with grease. If you coat the seal OD with sealant, be careful not to allow the sealant to contaminate the seal lips or clog the oil-return port.

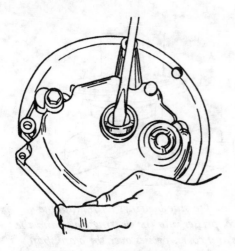

FIGURE 7-49. *Crankshaft seals come free with the help of a large screwdriver.* Tecumseh Products Co.

Installation is best done with a driver sized to match the OD of the rim (Fig. 7-50), although a piece of 2 × 4 works in an emergency. Drive the seal to the original depth (usually flush or slightly under-flush) unless the crankshaft exhibits wear from seal contact. In that case, adjust the seal depth to engage an unworn area on the crank, but do not block the oil port in the process.

The crankshaft must be taped during installation to protect seal lips from burrs, keyway edges, and threads. Cellophane tape, because it is thin, works best.

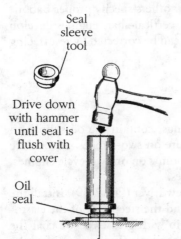

Seal sleeve tool

Drive down with hammer until seal is flush with cover

Oil seal

FIGURE 7-50. *The correctly sized driver confines installation stresses to the outer edge of the seal retainer.*

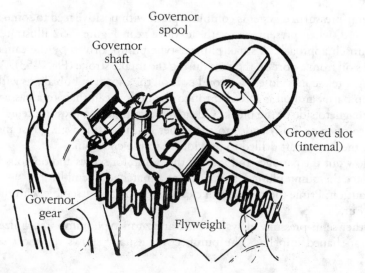

Governor
spool

Governor
shaft

Grooved slot
(internal)

Governor
gear

Flyweight

FIGURE 7-51. *Typical centrifugal governor. Flyweights react against a plastic spool.*

Governor

The unit shown in Figure 7-51 is typical of most centrifugal governor mechanisms. Paired flyweights, driven at some multiple of engine speed, pivot outward with increasing force as rpm increases. The spool translates this motion into vertical movement that appears as a restoring force on the carburetor throttle linkage.

Work the mechanism by hand, checking for ease of operation and obvious wear. The governor shaft presses into the block or flange casting; should it need replacement, secure the shaft with Loctite bearing mount and press it to the original height.

Lubrication systems

Lubrication systems require careful scrutiny. Conscientious mechanics will not release an engine unless all circuits have been traced, cleaned with rifle brushes, and buttoned up with new expansion plugs.

Any of three oiling systems are used. Most side-valve engines employ a dipper on the end of the connecting rod or a rotating slinger to splash oil about in the crankcase.

Semi-pressurized systems combine splash with positive feed to some bearings and, when present, to overhead-valve gear. Figure 7-52 illustrates the Tecumseh approach. A small plunger-type pump, driven by the camshaft, draws oil from a port on the cam during the intake stroke (Fig. 7-53). As the plunger telescopes closed, a second port on the camshaft hub aligns with the pump barrel and oil is forced through the hollow camshaft to a passage on the magneto side of the block. Oil then flows around a pressure relief valve (set to open at 7 psi) and into the upper main-bearing well. Most models have the crankshaft drilled to provide oil to the crankpin.

Blow out the passages and inspect the pump for scores and obvious wear. Replace the pump plunger and barrel as a matched assembly.

Caution: Prime the pump with clean motor oil and assemble with the flat side out.

Other semi-pressurized systems use an Eaton-type pump, recognized by its star-shaped impeller. The pump cover usually shows the most severe wear.

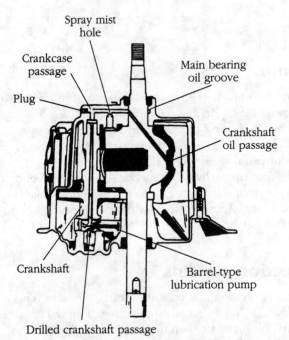

FIGURE 7-52. *A barrel-type pump supplies oil under pressure to the upper main bearing and crankpin on Tecumseh vertical-shaft engines. Other parts lubricate by splash.*

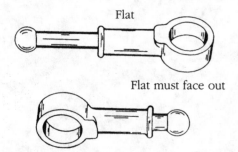

Flat

Flat must face out

FIGURE 7-53. *Plunger pump drives off a camshaft eccentric.*

Full-pressure systems deliver pressurized oil to all crucial bearing surfaces, although some parts receive lubrication from oil thrown off the crankpin and by oil flowing back to the sump. The Kohler system is typical (Fig. 7-54). A gear-type pump supplies oil to the pto-side main bearing, crankpin, and camshaft. The hollow camshaft carries oil to the magneto-side main bearing and crankpin. A pressure-relief valve limits pressure to 50 psi.

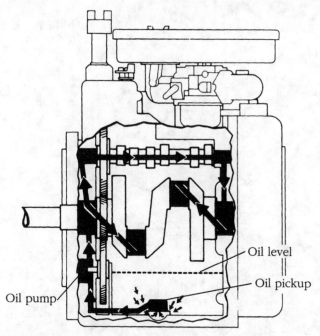

Oil level

Oil pickup

Oil pump

FIGURE 7-54. *Kohler full-pressure system utilizes a gear-driven pump and a hollow camshaft.*

Appendix
Internet resources

Exercise extreme caution when researching engine information on the Web. Sites that offer free service manuals or other documentation are sources of ransomware. Download material only from trusted factory websites.

Two- and four-stroke engines

Arrow

Technical tips and replacement parts for Arrow, Ajax, Fairbanks-Morse, Caterpillar, and Waukesha engines: http://www.arrowengine.com/en/. Some Arrow C-series oil-field engines have been in continuous operation for 75 years with only annual oil and spark-plug changes.

Briggs & Stratton

Free downloadable owner's and parts manuals, and printed shop manuals for purchase: https://shop.briggsandstratton.com/us/en/parts-and-accessories/repair-manuals.

Craftsman

Replacement parts and parts lists for equipment sold by Sears: www.searspartsdirect.com/. Owner's manuals: http://www.searspartsdirect.com/partsdirect/user-manuals.

Cummins Onan

The *RV Generator Handbook*, a very comprehensive owner's manual: https:// power.cummins.com/sites/default/files/literature/rv/F-1123-EN.pdf. Genset operator's manuals and spec sheets: http://power.cummins.com/specification -and-datasheets.

Dolmar

Owner's manuals and a small collection of training material: http://www .dolmarpowerproducts.com/know_how/training_material.

Echo

Parts catalogs, safety literature, and owner's manuals: http://www.echo-usa .com/Support-Help/Technical-Documents. Videos, some of which deal with service procedures: http://www.echo-usa.com/Videos/How-To-Videos/ECHO -Equipment-Preventative-Maintenance.

EFCO

Exploded-parts drawings for chainsaws and garden tools: http://www.efco power.com/resources/data-sheets.

Hitachi

Shop manuals available from factory authorized dealers or by calling 1-866-213-3373. For further information: http://www.hitachiconstruction.com/ service-support/order-manuals/.

Homelite

Operator's manuals and replacement parts: http://www.homelite.com/service_ support/parts_locator.

Honda

Owner's manuals: http://engines.honda.com/parts-and-support/owners-manuals/. Printed shop manuals can be obtained from:

- Honda Engines' eBay store
- Honda Engines' Amazon storefront
- Local Honda dealers

Husqvarna

Owner's manuals: http://www.husqvarna.com/us/support/manuals-downloads/.

John Deere

Shop and parts manuals for purchase: https://techpubs.deere.com/.

Jonsered

Operator's manuals and parts lists: http://www.jonsered.com/us/support/download-manuals/?query=Z46R&types=O.

Kawasaki

Owner's manuals: http://www.kawasakienginesusa.com/support/manuals. Shop manuals can be purchased from dealers.

Kohler

No-charge downloadable shop and owner's manuals: http://www.kohler engines.com/manuals/landing.htm.

Makita

Operator's manuals and parts lists: https://www.makitatools.com/products/details/.

McCulloch

How-to videos: http://www.mcculloch.com/int/support/how-to/.

Poulan

Operator's manuals: http://www.poulan.com/customer-support/. Replacement parts: http://www.ordertree.com/poulan/.

RedMax

Operator's manuals, data sheets, and parts manuals: http://www.redmax.com/.

Shindaiwa

Operator's and parts manuals: http://www.shindaiwa-usa.com/Tech-Support/Technical-Docs.aspx?s=EB802RT.

Stihl

Operator's manuals: https://www.stihlusa.com/manuals/instruction-manuals/. Downloadable, no-charge shop manuals: http://opeforum.com/threads/stihl-workshop-manuals.1112/.

Subaru

Operator's, parts, and shop manuals: http://www.subarupower.com/products/manuals/.

Tecumseh

Purchase printed shop manuals: http://www.tecumsehpartstore.com/. Basic troubleshooting and service information: http://www.tecumsehpower.com/CustomerService/BSI.pdf.

Wisconsin

Parts manuals: http://www.continentalengines.com/wisconsin-engines-parts.html. Provides downloadable parts manuals.

Fuel systems

Carburetor adjustment

https://www.youtube.com/watch?v=D7mnQKIVMXs.

Tillotson air-head carburetors

http://www.kartingmagazine.com/features/tillotson-diaphragm-carbs-for-karting/; http://www.tillotson.ie/techinfo-service.phpALL.

Walbro carburetor service manuals

http://www.walbro.com/service-manuals/; http://www.tanaka-usa.com/docs/service-documents/walbro_service_manual.pdf?sfvrsn=2.

Zama servicing

http://www.zamacarb.com/page/disassembly_servicing_1.

Zama and Walbro tools

http://www.amazon.com.

Ignition systems

Spark plugs

http://www.autolite.com/produ... ¡awn-and-garden.aspx#xtreme_start_spark_plug;
https://www.ngk.de/en/technology-in-detail/spark-plugs/diagnosis/defects/;
https://www.ngk.de/en/technology-in-detail/spark-plugs/thermal-behaviour/.

Alternative ignition systems

http://gardentractorpullingtips.com/ignition.htm.

Forums

Do-It-Yourself Forum, outdoor power equipment repair: http://www.doityourself.com/forum/outdoor-gasoline-electric-powered-equipment-small-engines-70/.

Hobby Talk Forum, two- and four-stroke garden equipment repair and small-engine swaps: http://www.hobbytalk.com.

My Lawnmower Forum, riding- and push-mower service: http://www.mylawnmowerforum.com/forum/.

PER Small Engine Forum, 12,000-plus small-engine posts: http://per/com/forum/.

Index

Note: Page numbers followed by *f* indicate figures; page numbers followed by *t* indicate tables.

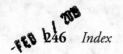